ENVIRONMENT, POWER, AND SOCIETY

W8 - AEH -763

ENVIRONMENT, POWER, AND SOCIETY

HOWARD T. ODUM

University of North Carolina
Chapel Hill

WILEY-INTERSCIENCE

A Division of John Wiley & Sons, Inc.

New York   London   Sydney   Toronto

Library of Congress Catalog Card Number: 78-129660
Cloth: ISBN 0 471 65270 9
Paper: ISBN 0 471 65275 X
Printed in the United States of America

10 9 8 7 6 5 4 3 2 1

To Howard Washington Odum

PREFACE

In recent years studies of the energetics of ecological systems have suggested general means for applying basic laws of energy and matter to the complex systems of nature and man. In this book, energy language is used to consider the pressing problem of survival in our time—the partnership of man in nature. An effort is made to show that energy analysis can help answer many of the questions of economics, law, and religion, already stated in other languages. Models for the analysis of a system are made by recognizing major divisions whose causal relationships are indicated by the pathways of interchange of energy and work. Then simulation allows the model's performance to be tested against the performance of the real system.

When systems are considered in energy terms, some of the bewildering complexity of our world disappears; situations of many types and sizes turn out to be special cases of relatively few basic types. An ideal expressed by the Society for General Systems Theory and elsewhere suggests that a general systems view of the world is possible and preferable in the orientation and education of man. Toward this ideal energy flows are illustrated with ecological systems and then applied to all kinds of situations from very small biochemical processes to the large overall systems of man and the biosphere. Energy diagraming helps us consider the great problems of power, pollution, population, food, and war free from our fetters of indoctrination.

Intended for the general reader, this account also attempts to introduce ecology through the energy language. Hopefully, it may be useful to the widespread efforts under way in undergraduate colleges to develop courses in human ecology that are pertinent to a new generation grappling for survival on the planet. Whereas mathematics is not required, footnotes and an appendix do provide formulations of some basic definitions emphasizing that the concepts in use are all quantitatively defined and measurable.

Grateful acknowledgments are due my former professor, George Evelyn Hutchinson of Yale University, for his example and my wife, Virginia Wood Odum, for encouragement to venture in interscience generalization. Many colleagues and former students here and at my former universities (Florida, Duke, Texas, Puerto Rico) aided in development of these theses. A memorial tribute is due Richard C. Pinkerton, whose untimely death cut short an earlier collaboration. Research on ecological models and microcosms was supported by the U. S. Atomic Energy Commission and the National Science Foundation. The book is dedicated to my father Howard Washington Odum who suggested a synthesis of science and society. I acknowledge the shared effort toward this aim with my brother Eugene P. Odum, University of Georgia.

This work was completed while at North Carolina. My present address is Graduate Research Professor, Department of Environmental Engineering, University of Florida, Gainesville, Florida.

HOWARD T. ODUM

Chapel Hill, North Carolina
May 1970

CONTENTS

ENVIRONMENT, POWER, AND SOCIETY

1

THIS WORLD SYSTEM

This book is about nature and man. In its structure and function, nature consists of animals, plants, microorganisms, and human societies. These living parts are in turn joined by invisible pathways over which pass chemical materials that cycle round and round being used and reused (Fig. 1–1(a) and (b)) and over which flow potential energies that cannot be reused (Fig. 1–1(c) and (d)). The network of these pathways forms an organized system from the parts. In addition, the more complex systems for self-control have specialized communication circuits such as the behavioral cues passing between animals. In the parts of the system involving human interchange, there are special kinds of information exchange, such as human language, and special units of economic exchange such as dollars. A study of nature and man is thus a study of systems. The budget of fuel energy received by the systems of nature controls the amount of structure that can exist and the speed at which processes can function. The small ecological systems, the large panoramas that include civilized man, and the whole biosphere of the planet earth—all receive only certain amounts of energy. Hence we approach man and nature by studying the limited energy of environmental systems.

MAN WITHIN NATURE'S SOLAR PATTERN

When human societies first evolved as a significant part of the systems of nature, man had to adapt to the food and fuel energy flows available to him, developing the now-familiar patterns of human culture. Ethics, folkways, mores, religious teachings, and social psychology guided the individual in his participation in the group and provided means for using energy sources effectively. The energy source was sunlight which is

1

spread out so evenly that it is not directly available to man until after some has been concentrated and much has been necessarily lost by the plants and animals. Societies that were able to survive had to gather food and distribute energies within the social system for their successful continuance, and they developed the group organization necessary for these purposes. The social systems adapted to meet changing conditions such

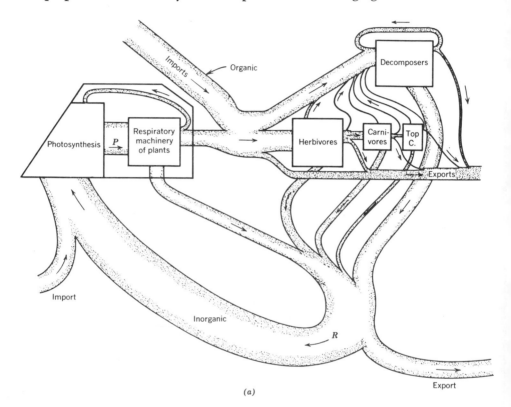

(a)

Figure 1–1 Flows of material and energy in environmental systems, which have inputs and outflows of materials and energy. P is gross photosynthesis and R total respiratory consumption. (a) Mineral flows and cycles through plants and animals. (b) Abbreviated diagram of material flows; carbon in a system of underwater meadows, Silver Springs, Florida [10]. (c) Flows of sunlight and food into an ecosystem with the same categories as shown in (a). (d) Abbreviated energy diagram of the ecosystem at Silver Springs, Florida, with categories as in (b). Energy from physical stirring is the fall of water level within the system. (Energy value of chemical salts in the flowing water was calculated using the expression $\Delta F = RT \ln C_2/C_1$. C_1 is the concentration in the inflowing water; C_2 for the flow into plants is the concentration of water leaving the ecosystem; C_2 for the outflowing water is that of the general nutrient level of the region's waters.) See Odum [10].

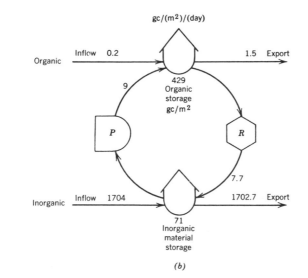

gc/(m²)/(day)

Organic   Inflow 0.2 →   1.5 Export →

9   429
           Organic
           storage
           gc/m²

P   R

7.7

Inorganic Inflow 1704 →   1702.7 Export →

71
Inorganic
material
storage

(b)

Sunlight

Imports

Photosynthesis   P   Respiratory machinery of plants

Decomposers

Herbivores   Carnivores   Top C.

Exports

R

Heat

(c)

3

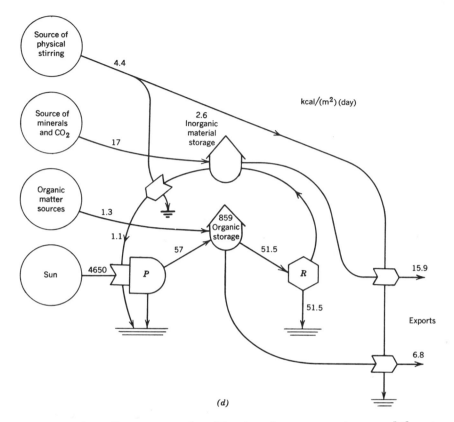

(d)

as overcrowding, fluctuations of yield, crises from competitors, and threats from internal disorder. The old pattern is shown in Figure 1–2(a).

One self-stimulating principle of the primitive group was to allocate control of the energy flow of the group to individuals in proportion to the work they did to increase that flow. Such energy rewards took various forms, such as control of property, political power, and status influence. The economic system was simple, and economic reward often reflected the energy control gained.

In these earlier times, when man's simple role in the environmental system was based on sunlight energy, plant production was very important. Since plant production depended on area, land use was a principal concern of the social system. But only a small part of the earth's surface and the energy flows of the environment was really controlled by man, and the checks and balances of the natural systems prevailed. Man did survive with these regimes, however, and his social processes for decision

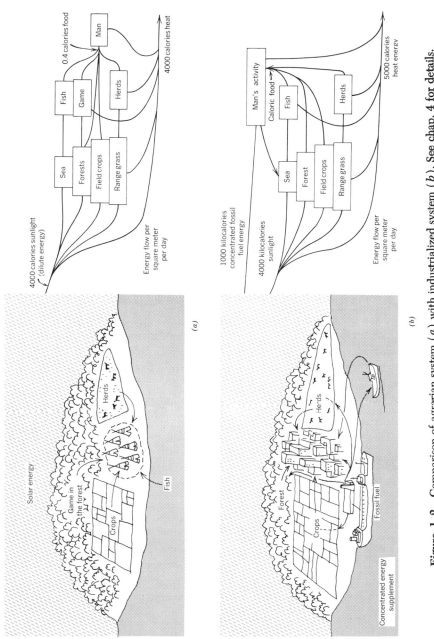

Figure 1-2 Comparison of agrarian system (*a*) with industrialized system (*b*). See chap. 4 for details.

5

were developed and brought to modern times on the basis of suffered protection by the natural systems.

MAN TAKES OVER NATURE WITH FOSSIL FUELS

In modern times man's energetic relation to his environment has changed, for he now has much more fuel under his control, and it enters through a different route. Man's new industrialized system now derives its energies from the flows of concentrated fossil fuels, coal, and oil. Much of this energy flow is put back into his environmental system so that his yield of food and critical materials is greater. The new energy flow pattern is illustrated in Figure 1–2(b).

The changes have come so fast that man's customs, mores, ethics, and religious patterns may not have adapted to them. We may wonder whether the programs controlling the behavior of the individual human being, whether he is a leader or a person led, are realistic in providing for the best adaptation to the new systems of energy flow. Man's role in the environment is becoming so enormous that his energetic capacity to hurt himself by upsetting the environmental system is increasing. His system is becoming so large and complex and is changing so rapidly that it is more and more difficult to learn of the patterns in his own fuel supply. Money, energy flows, and causal actions can no longer be easily equated. The accumulation point of money in the network may be much removed from the origin of the work done, of the energy inflow, or of the forces causing stress.

Only a small part of the total controlled energy is now processed by the individual person or by work that is recognizably personal. More and more of the energy flows are in machines of the system. We may wonder whether the individual human being understands the real source of the bounty to him of the new energy support. How many persons know that the prosperity of some modern cultures stems from the great flux of oil fuel energies pouring through machinery and not from some necessary and virtuous properties of human dedication and political designs? It is not easy to fit into our thinking the concept that part of the energy for growing potatoes comes from fossil fuel. In the new system much of the higher agricultural yields are only feasible because fossil fuels are put back into the farms through the use of industrial equipment, industrially manufactured chemicals, and plant varieties kept in adaptation by armies of agricultural specialists supported on the fossil-fuel-based economy.

How many of the old landmarks in the relation of man and energy are gone? How often now is the economic purchasing power supplied a man given in relation to the work he processes? How often now is the energy processed by a group in proportion to the land area of forests and fields controlled by that group? How often now do the instructions on good and bad behavior given to the young have convincing reality for the best energetic flows of the system or the individual?

In a thousand ways, the accelerating changes accompanying an increasing budget of energy for man raise questions of his ultimate role and survival. The pessimist talks of man as the next in line of the extinct dinosaurs and other predominant types that once inherited the earth. Like the dinosaurs, man is developing a system of specialists and giant mechanisms. The extinction of the dinosaurs serves as a warning that our so-called progress may not be a safe plan for survival.

As the industrialized areas increase, how much longer will the biological cycles in the uninhabited environments be able to absorb and regenerate the wastes and thus prevent self-poisoning of waters and atmosphere? As man's system becomes large enough to control and prevail in the flows of the biosphere, will he understand it well enough to prevent disaster? If his energy sources begin to decline, can he return to a minor energetic position in the earth system without a collapse and extinction of culture as we now know it?

Critical issues in public and political affairs of human society ultimately have an energetic basis, and the increasing number of urgent energy demands measure the accelerating changes in the system containing man. The action programs needed on such public issues as birth control, land ownership, man in space, war prevention costs, nuclear power, zoning of land space, human medical maintenance, and world economics must each be limited and fitted into the overall energetic budget for a successful system of nature. Energetic budgeting in simpler systems of nature that do not include man is readily discernible and has often been described. The same principles apply to the vast industrialized system. Yet how seldom is the energetic budget discussed or its rules applied to human society. Subconscious attempts to use economic data as a substitute for energy data may be misleading.

The energetic processes in a dense reef of oysters in a bay on the Gulf Coast of the United States are basically the same as those of an industrial city (see the energy diagrams in Fig. 1–3). The fossil record too is beautiful with its magnificent chronology of ancient reef systems. But the energy networks of these mineral reef systems are gone and with them the component species that dominated. What about the future of man's industrial reefs?

Man's survival will probably depend on his being able to see what his

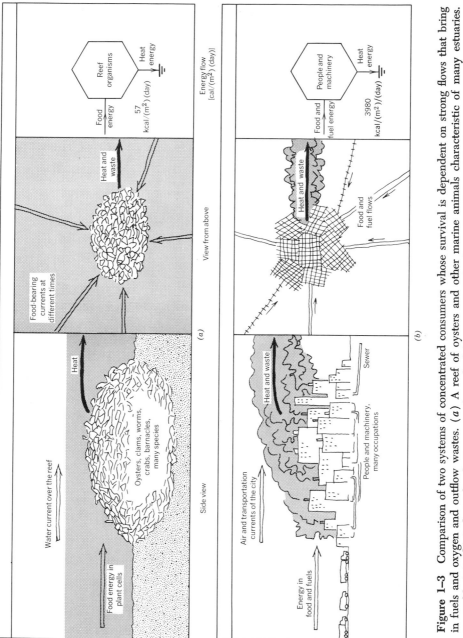

Figure 1-3 Comparison of two systems of concentrated consumers whose survival is dependent on strong flows that bring in fuels and oxygen and outflow wastes. (*a*) A reef of oysters and other marine animals characteristic of many estuaries. See Table 2.2. (*b*) Industrialized city; see Wolman [16] for account of the flows of a city's metabolism.

vast human system has become in relation to preceding and possible earth systems. And he must acquire the necessary understanding rapidly enough to adapt his opinions, folkways, mores, and action programs to the great new systems and provide a continuing survival path for them. Since decisions on such matters in the arena of public affairs are ultimately made according to the beliefs of the citizens, it is the citizens who must somehow include the energetics of systems in their education. In some way the behavior of the large and small systems must be understood and that knowledge must be communicated to the dispersed intelligence of the modern decision apparatus.

Can we learn the nature of macroscopic environmental systems including man, and can we clarify their functions so that they are visible to the citizen and his leaders? Is there an adequate old or new morality which is preadapted to guide individual actions for the survival of the group in the realities of the new energetic situations?

Investigators with many kinds of training and background are approaching these systems problems, some concentrating on human relations systems alone, some concentrating on simple ecological models like the reef which do not include man, some trying to develop general systems theory, and some working on the electrical computer models. Let us follow the principles of energy in environmental-scale systems that include our industrial civilization. To see these patterns which are bigger than ourselves, let us take a special view through the macroscope.

THE MACROSCOPIC SYSTEMS POINT OF VIEW

In the 1600s when Leeuwenhock ushered in the centuries to be enlightened by the study of the invisible world with the microscope, and when some of the atomistic theories of the Greeks received step-by-step observable verification in chemical studies, concepts of the structure and function of the natural world emerged as parts within parts within parts. Many of the advances of mankind have come from these microscopic dissections. Yet in the twentieth century the ever-accelerating knowledge of the microscopic view has not provided us with the solution of some classes of problems concerned with man's environment, his social systems, his economics, and his survival, for the missing information is not wholly in the microscopic components or in identification of the parts. Man now sees the world of parts well indeed. But he is only beginning to see the systems of which he is one of the parts.

Whereas two centuries of scientific progress were derived from the microscopic work, we find the contemporary world beginning to look through a macroscope of systems science and acquiring ways to discern the broad features and mechanisms of a system of parts. Bit by bit the machinery of the macroscope is evolving in various sciences and in the philosophic attitudes of students. The daily maps of worldwide weather, the information received from the high-flying satellites, the macroeconomic statistical summaries of nations and the world, the combined efforts of international geophysical collaborations, and the radioactive studies of cycling chemicals in the great oceans all stimulate the new view. Whereas men used to search among the parts to find mechanistic explanations, the macroscopic view is the reverse. Men, already having a clear view of the parts in their fantastically complex detail, must somehow get away, rise above, step back, group parts, simplify concepts, interpose frosted glass, and thus somehow see the big patterns. Astronomical systems, although infinitely larger, are seen through such distances that only the main features show; on earth progress is slow because we are too close to see. As in the old adage about the forest and the trees, we cannot see the pattern for the parts. Figure 1–4 is a cartoon view of the macroscope, showing the steps we must take in going from detailed data surveys to system viewing and prediction.

In learning how to build from parts into larger wholes and patterns,

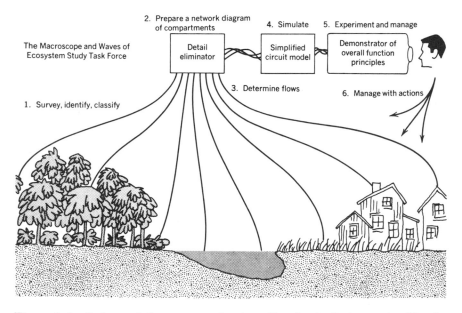

Figure 1–4 Cartoon of the macroscopic view. The detail eliminator simplifies by grouping parts into compartments of similar function.

certain sciences are finding new and clearer lenses for the macroscope. Some of these are concepts from the general network of science and pertinent mathematics. The building of electronic systems models is an example of the joining of well-understood parts to comprehend the group phenomena. Other examples are the computer models of the world's carbon cycle or radioactivities.

In this book, therefore, the reader is invited to view the world and man through the macroscope. We hope eventually to discern the great clanking wheels of the machinery in which man is such a small component. And perhaps in the end we can predict the times and places in which man's services will continue to be used or guess whether and when other parts will replace him in the drama of the earth.

THE BIOSPHERE

We can begin a systems view of the earth through the macroscope of the astronaut high above the earth. From an orbiting satellite, the earth's living zone appears to be very simple. The thin water- and air-bathed shell covering the earth—the biosphere—is bounded on the inside by dense solids and on the outside by the near vacuum of outer space. Past the orbiting capsule radiant energy from the sun enters the biosphere, and eventually equal amounts pass outward as flows of heat radiation. In the haze of height a few grand phenomena of the energy flows can be observed. There is suggestion of a great sheet of chlorophyll, there are cyclones, and there are characteristic cloud forms of the weather belts, but the miraculously cascading machinery of parts within parts within parts is not even visible. From the heavens it is easy to talk of gaseous balances, energy budgets per million years, and the magnificent simplicity of the overall metabolism of the earth's thin outer shell. With the exception of energy flow, the biosphere can be seen as the closed system that it for the most part is (see Fig. 1–5), a closed system of the type whose materials are cycled and reused.

The biosphere is the largest ecosystem, but the forests, the seas, and the great cities are systems also. The great chunks of nature also have subsections and zones which are organized by their physical processes and organisms into systems of function. Large and small parts operate on their budgets of energy, and what can and cannot be done is determined by the simple laws that govern the system. Any phenomenon is controlled both by the working of its smaller parts and by its role in the larger system of which it is a division.

Among the subsystems are some based on the concentrated fuel flows

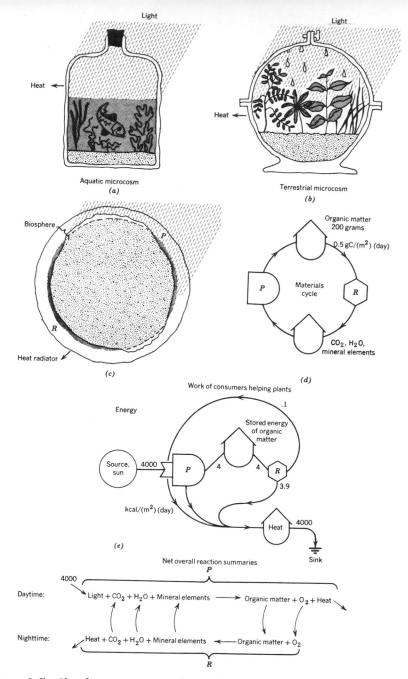

Figure 1–5 Closed systems supported entirely by sunlight and their overall reactions. (*a*) Aquatic closed system; (*b*) Terrestrial microcosm; (*c*) Biosphere; (*d*) Cycle of materials between production (*P*) and respiratory consumption processes (*R*); (*e*) Energy diagram for the systems: (*a*)–(*c*).

12

that man has the power to control. Elsewhere man participates in systems in which the energy at his disposal is too small to regulate. Because man's power is still insufficient to control the overall biosphere, he is protected in his ignorance by the great stabilizing storages of the oceans and atmosphere. An accelerating growth of energy flux in man's system, however, is dangerously outdistancing his knowledge of its workings.

In the astronaut's view from the heavens, how are the large and small systems structured, how do they function, and how are they organized with larger wholes? What principles are recognizable in the grand view, in the subsystems, and in the systems of man? What is the common functional order of the macroscopic world?

The Living Fires of the Earth

With the turning of the earth, the sun comes up on fields, forests, and fjords of the biosphere, and everywhere within the light there is a great breath as tons upon tons of oxygen are released from the living photochemical surfaces of green plants which are becoming charged with food storages by the onrush of solar photons. Then when the sun passes in shadows before the night, there is a great exhalation as the oxygen (O_2) is burned and carbon dioxide (CO_2) pours out, the net result of the maintenance activity of the living machinery. During the day while the oxygen is generated, a great sheet of new chemical potential energy in the form of new organic matter lies newborn about the earth, but as the oxygen is consumed in the darkness, the organic matter disappears like firewood in a bonfire and releases heat through the night.

The living process associated with the cover of green chlorophyll during the lighted period in forests, lakes, oceans, and deserts is called photosynthesis by those who study a small segment of it and primary production by those who consider great masses of it. Because various kinds of living communities have been studied in detail, some estimates can be made of the contribution to the whole system per day. The letter P (for production) is the abbreviation given to the process (Fig. 1–5).

The living fire that dominates the dark night period but is masked by primary production during the day is called respiration by those who study segments of it; more often it is called food and fuel consumption by those who consider large amounts. We can make estimates of the consumption in the biosphere. In the overall equation for consumption, materials are returned to the inorganic state, ready for primary production again. The letter R is the abbreviation for return respiratory consumption flux. Together P and R function simultaneously or consecutively to provide a cycle of chemical elements. Experiments with closed microcosms

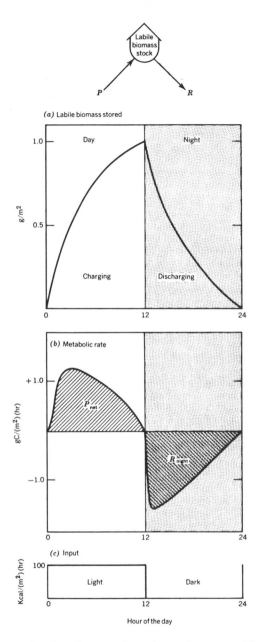

Figure 1–6 Diurnal pulse of power flows in a microcosm with P and R processes graphed [1]. (*a*) Storage of organic matter during periods of charge in the light and discharge in the dark. (*b*) Rates of photosynthesis and respiration. (*c*) Inflowing light energy.

14

(Fig. 1–6) show a regular day and night rise and fall of the carbon dioxide used by production and released by respiration. The chemical reaction is similar to that accomplished by fire, except that the potential energy is passed into various kinds of useful work in the cold ecological processes whereas in fire it is dispersed as useless activity of the atmosphere.

On the average P and R tend to be equal in a system closed to matter. In summer P tends to exceed R but in winter R tends to exceed P. The carbon dioxide in the air diminishes in the spring and increases in the fall [3].

The biospheric averages of P and R have been estimated as about one gram of organic matter per square meter per day. This is about four kilocalories of potential energy stored daily as organic matter and burned again.

The Coupling Interchange of Two Main Energy Flows

About half of the energy that comes to the earth from the sun is in wavelengths incapable of directly driving photosynthesis and is absorbed in various materials of the biospheric surface, becoming heat. This directly absorbed heat, plus the heat released from photosynthetic machinery using the other wavelengths, plus heat released when organic matter is consumed all add together to cause higher temperatures near the equator and lower temperatures in shaded regions. Between high- and low-temperature zones a giant heat engine operates, creating the great wind and water current systems of the earth. The energies processed by the earth's heat engine contribute to the photosynthetic and respiratory process by bringing raw materials such as rain and fertilizer minerals more rapidly to the sites of production; they also aid some food chains by moving organic matter to the sites of consumption. Hence the P and R processes of the biosphere are closely linked energetically with the earth's physical processes, its heat, winds, and water currents. For example, see Fig.1–1(d).

Elements That Cycle Together

The living processes require a flow of the materials of life such as carbon. Oxygen, hydrogen, and mineral elements are shown in the equations of Figure 1–5. Each element flows in pathways guided by the living organisms. We can think of these pathways as circuits of the ecological system. When we write a carbon circuit for a general ecosystem, we are at the same time writing something of a first approximation for other elements such as oxygen, phosphorus, and nitrogen that have rather similar ratios

to carbon throughout the system, especially in the total effect of the consumers regenerating for the producers as shown in Figure 1–5. The
knowledge that elements occur in characteristic ratios to carbon makes it
possible to calculate their flows from data on carbon metabolism. Carbon
is the best element to use for overall material calculations (if only one is
to be used) because it is the largest fraction, 40 to 80 percent of organic
matter and fuels. The ratio of an element used to carbon processed is
called a metabolic quotient.

There are metabolic quotients characteristic of each type of system.
Many pollution problems can be generalized as a change in element ratios
in the cycles through the insertion of waste flows from some industry of
man. By and large man's culture has a characteristic mix (ratio of elements). The North Atlantic Ocean also has a certain mix as Redfield
[13] showed in extending to nitrogen and phosphorus the physiological
concept of respiratory quotients (oxygen to carbon ratios). The innova-
tive principle stated by Redfield was that stable regenerative zones cycle
elements in the same ratio as the productive zone. Table 1–1 shows mixes
for man's culture as taken from his sewer and for an aquatic system. The
wastes of the human system are quite different and would change the
other system if mixed.

Another property common to the system and the subsystems is the
storage of both inorganic materials and the organic quantities living and
dead (see Fig. 1–1). In a later chapter we shall discuss these storages
as means for smoothing out fluctuations in input. The relation of P, R,
the organic storage, and the inorganic mix in matter-closed systems is
portrayed in Figure 1–5. Shown in Figure 1–1(b) is a system with four
exchanges with the outside (organic import and export and inorganic
import and export). The relative flux of the four boundary flows varies
from system to system. For a steady state the sum of the inflows must
equal the outflows. A steady state is one that holds a constant pattern of
flows, cycles, storages, and structures.

THE CHANGING METABOLIC ROLE OF MAN

Figure 1–2 shows something of the contrast between the new industrial
system and the old agrarian ways of man, emphasizing the different
effects on all processes of a diffuse solar energy source prorated broadly
on an area of agricultural land and of a highly concentrated fossil fuel
distributed to an industrial city. As shown in Table 2–2, the concentration of metabolism in an industrialized center becomes very large in

Table 1-1   Chemical Ratios of Regenerative Cycles[a]

Element	Temperate Ocean[b]	Human Association[c]	World Agriculture
Total dry weight of food used	2.3	1.8	2[d]
Oxygen consumed in respiring food	1.9	2.9	—
Carbon respired and released in waste	1	1	1[d]
Nitrogen processed	0.18	0.2	0.029[e]
Phosphorus processed	0.024	0.016	0.011[e]

[a] Ratios to carbon as one.
[b] Sverdrup, Johnson, and Fleming [14].
[c] Per capita wastes, Meyers [8]: carbon, 296 g/day; oxygen, 860 g/day; food 520 g/day; Fair, Geyer, and Okun [2]: N, 7.9 g; P, 1 g. W. Stumm and J. J. Morgan (C:N:P 60:12:1).
[d] See Table 2-2.
[e] Fertilizer production, 1963; Nitrogen, 15.2 million metric tons per year; phosphorus, 5.7 million metric tons per year (McCune, Hignett, and Douglas [6]).

relation to that of the system formerly occupying the space. Such concentration of consumers also occurs among other natural groupings, for example, the consumer reef in the sea. See Fig. 1-3.

In the agrarian system with man and his animals living off the land, there was often a balance of primary production (P) and total consumption (R) in the course of the year. Man's net annual effect on the gases of the atmosphere, on the concentration of minerals, or on the future was small, for the system was balanced as an aquarium is balanced. As agrarian systems multiplied the world over, the biosphere also remained approximately the same. Some organic matter was stored as fuel and oil, but the amount per year was tiny indeed.

In the industrial system with man living off a fuel, he manages all his affairs with industrial machinery, all parts of which are metabolically consumers. Even agriculture is dominated by machinery and industries supplying equipment, poisons, varieties, and services. The system of man has consumption in excess of production. The products of respiration—carbon dioxide, metabolic water, and mineralized inorganic wastes—are discharged in rates in excess of their incorporation into organic matter by photosynthesis. If the industrialized urban system were enclosed in a chamber with only the air above it at the time, it would quickly exhaust its oxygen, be stifled with waste, and destroy itself since it does not have the recycling pattern of the agrarian system. The problems with life support in 1970 on the space flight of Apollo 13 dramatized this principle to the world.

Industrialized lands are still a small portion of the world's surface compared to the forest, ocean, and agricultural areas, although the photosynthetic activity on land may have declined somewhat because agricultural areas are covered with green crops only part of the year. When we add up the balance sheet for the biosphere including both the areas of the world, we find that human industries add about a 5 percent excess of consumption over production. Presumably the rising concentrations of carbon dioxide and mineral nutrients released by consumption have also accelerated excess production in enough places to bring about a new overall balance in the biosphere at the higher level of carbon dioxide concentration. As long as the industrialized areas are in small ratio to the agrarian areas, the biosphere can remain balanced without a major effort to increase plant production.

The biosphere with industrial man suddenly added is like a balanced aquarium into which large animals are introduced. Consumption temporarily exceeds production, the balance is upset, the products of respiration accumulate, and the fuels for consumption become scarcer and scarcer until production is sufficiently accelerated and respiration is balanced. In some experimental systems balance is achieved only after the large consumers which originally started the imbalance are dead. Will this happen to man?

CIRCUIT DIAGRAMS OF THE SYSTEM'S MATERIALS

One way to predict the consequences of change in a system is to draw a systems diagram as a model, including values for flows and storages. Then the relative magnitude of the change can be compared with the normal. Such approaches, which relate the parts and configurations of their connections to performance, are in the realm of the science of systems. Systems principles may be introduced with a materials circuit diagram for carbon in the biosphere (Fig. 1–5(d)).

The circuit diagrams in Figures 1–1(b) (d) include the main features of a system: inflow of organic fuel, outflow of organic matter, inflow of inorganic raw materials, outflow of inorganic wastes, the general process of photosynthesis (P) creating organic matter and potential energy from inorganic matter by degradation of light, and the respiration (R) of the system representing all the work driven by the combustion of organic fuel. Included in the work done in respiration are the orderly arrangements and replacements that are part of maintenance of the structure of the system. Also included are the flows in and out of storage reservoirs of inorganic and organic matter. In this system the outflows are in proportion to the concentrations stored. Numbers indicate the rates and the quantities stored. Both structure and functions are shown. Some idea of the time required for turnover is found by dividing quantity stored by rates at which materials pass through storage. The diagrams illustrate the constraining principle that there be conservation of mass. In the system closed to matter (Fig. 1–5), there is no inflow and outflow and the parts are interdependently linked by the internal cycle.

THE STABILIZING PROPERTY OF A LOOP

Configurations in a system determine its behavior following change. The main cycles of materials such as carbon have a regulatory effect which somewhat ameliorates changes.

Materials passing from one group of processes (P) to the other (R) provide a self-rewarding loop (Fig. 1–5(d)) with each process stimulating the next. If one of the two processes should slow down, so would the other until storage concentrations build up and force the process back to its regular rate. The rise and fall of storages and rates in microcosms for day and night are drawn in graph form in Fig. 1–6. [Flow is proportional to the quantity stored.] Rates increase with storages.[1]

These mechanisms of circular stimulation are readily observed in microcosms; we find that the rate of photosynthesis during the day depends on the amount of respiration of the previous night and vice versa. In a general way the earth biosphere apparently works by circular stimulation too, even though longer time periods are involved. When a summer regime running on a pulse of light alternates with a winter regime running on stored food, adequate quantities must be stored for all the metabolizing parts. When energy interruptions are long, larger quantities must be storaged, but when light regimes are almost continuous, amounts of food stored need not be so large.

For the whole biosphere the oxygen is stabilized by large storages in

[1] Semiclosed aquarium and terrarium systems such as those found in schoolrooms were studied for their carbon dioxide metabolism patterns after they had become stabilized and their kinetic properties were analyzed as feedback systems (Odum, Beyers, and Armstrong [11]; Milsum [7]; Odum, Lugo, and Burns [12]).

the air; carbon is stabilized by large storages in the sea and in limestones. By these examples we see that the properties of the whole system determine the outcome of flows and storages which we observe. A study of systems in this way is a part of the macroscopic view.

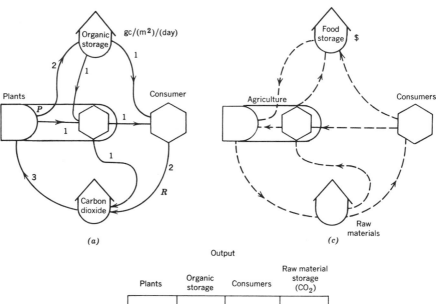

Figure 1–7 Rates of steady-state flow of carbon in a microcosm shown in network diagram (a) and in the form of a matrix diagram (b). Numbers are grams of carbon per square meter per day. The absence of a pathway in the network diagram is a zero in the matrix. In (c) a comparable economic system of man is shown with currency flowing in opposite direction from the carbon. See chap. 6.

MATRIX REPRESENTATION OF THE NETWORK DIAGRAM

Diagrams of mineral flows and energy flows are the main language used in this book to present and discuss the systems of nature and man. Any network diagram can be just as easily represented in a tabular form in which the separate units of the network appear as headings and the numbers in the table represent the energy flows or mathematical terms for that flow. A table relating parts by some system of relationship is a matrix. Everyone is used to such matrices in everyday life as a mileage table with cities as the headings and mileages between the cities indicated at the intersections (boxes).

In Figure 1–7 is an example describing the carbon cycle in both network and matrix form. From a microcosm study there are the same four units already introduced in the closed system diagrams in Figure 1–5(*d*), but more of the possible pathways are drawn. Representing systems in this way is a first step in the use of techniques of matrix algebra by systems analysts. A pertinent step after description as a matrix is to count the possible relationships that may have been overlooked. (There are systems known that have the pathways omitted in Fig. 1–7.) The number of possible boxes and the number of the boxes actually utilized in the system are properties of the complexity of the network. Network complexity is a property much used and studied in current research on all kinds of systems.

NEED FOR A NETWORK LANGUAGE FOR POWER

Although pathways, flow rates, and other descriptive properties of an environmental system can be represented with simple diagrams of material flow, such as carbon, only part of the whole environmental system is thus represented, only one type of flow can be given at a time, and only the processes affecting that material appear.

To understand a whole system and the full interaction of the parts, we must use a common denominator that expresses all the flows and processes together. Power is a common denominator to all processes and materials. If some portrayal of causal action is needed, the network diagrams must show the flow of causal forces. Since forces are generated from energetic storages, their lines of action may also be represented by the same lines that indicate energy delivery (Figs. 1–1(*d*), 1–2, and 1–5(*e*)). If potential sources of power deliveries are to be shown, the energy storages must also be given. Therefore in the next chapter we

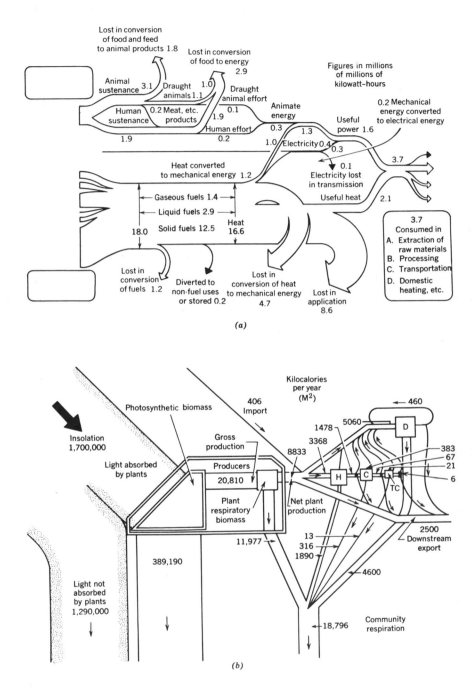

(a)

(b)

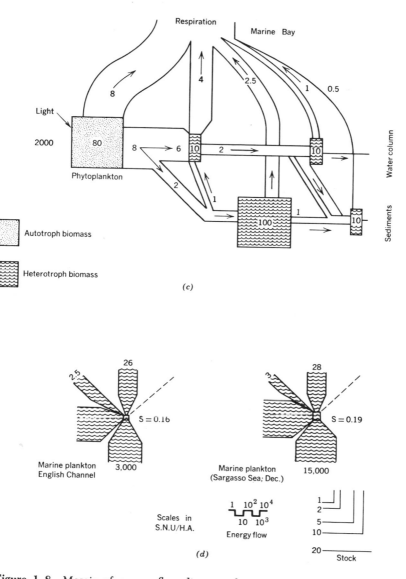

A Y-shaped energy flow diagram

Respiration

Marine Bay

Light

2000

80

Phytoplankton

Water column

Sediments

Autotroph biomass

Heterotroph biomass

(c)

26

2.5

S = 0.16

28

3

S = 0.19

Marine plankton
English Channel

3,000

Marine plankton
(Sargasso Sea; Dec.)

15,000

Scales in
S.N.U/H.A.

$1 \quad 10^2 \ 10^4$
$10 \quad 10^3$
Energy flow

1
2
5
10
20
Stock

(d)

Figure 1–8 Mosaic of energy flow diagrams by various authors and from various fields showing different styles and languages used. (*a*) World, Guyol [15]; (*b*) Silver Springs, Florida, Odum [10]; (*c*) English Channel, Odum [9]; (*d*) Marine Systems, MacFadyen [5]; (*e*) Data for Cedar Bog Lake, Minn., Kormondy [4]; (*f*) Forests, Woodwell and Sparrow [17].

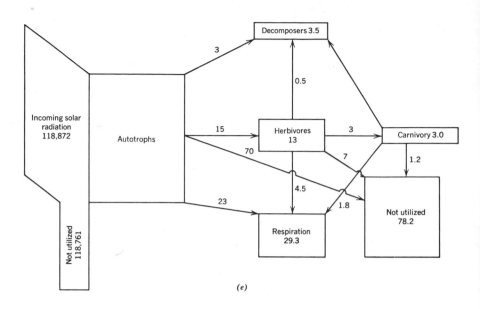

Decomposers 3.5

Incoming solar radiation 118,872

Autotrophs

Not utilized 118,761

3

0.5

15

Herbivores 13

3

Carnivory 3.0

70

7

1.2

4.5

23

1.8

Not utilized 78.2

Respiration 29.3

(e)

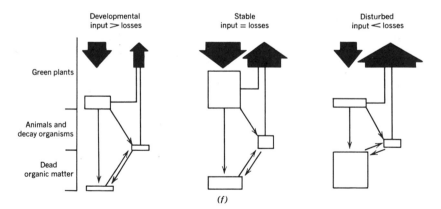

Developmental
input > losses

Stable
input = losses

Disturbed
input < losses

Green plants

Animals and
decay organisms

Dead
organic matter

(f)

describe some principles of energetics and an energy network language for the further consideration of the systems of man and nature. Power diagrams have been important in describing many systems, but there has not been any common convention in the use of symbols and diagrams. Samples of power diagrams are given in Figure 1–8.

24

REFERENCES

[1] Beyers, R. J., "The Metabolism of Twelve Aquatic Laboratory Microecosystems," *Ecol. Monogr.*, **33**, 281–306 (1963); A Characteristic Diurnal Pattern of Balanced Aquatic Microcosms. Publ. Inst., Marine Science, Texas, **9**, 19–27.

[2] Fair, G. M., J. C. Geyer, and D. A. Okun, *Water and Wastewater Engineering*, Vol. 2, Wiley, N.Y., 1968.

[3] Keeling, "Concentration and Isotopic Abundance of Atmospheric Carbon Dioxide," *Tellus*, **12**, 200–204 (1960).

[4] Kormondy, E. J., *Concepts of Ecology*, Prentice Hall, Englewood Cliffs, N.J., 1969.

[5] MacFadyen, A., "Energy Flow in Ecosystems and Its Exploitation by Grazing," in D. J. Crisp (ed.), *Grazing in Terrestrial and Marine Environments*, Blackwell, Oxford, 1964, pp. 3–20.

[6] McCune, D. L., T. P. Hignett, and J. R. Douglas, "Estimated World Fertilizer Production Capacity as Related to Future Needs," Tennessee Valley Authority, National Fertilized Development Center Muscle Shoals, Ala., 1966.

[7] Milsum, J. H., *Biological Control Systems Analysis*, McGraw-Hill, New York, 1966.

[8] Meyers, J., "Introductory Remarks to a Symposium on Space Biology: Ecological Aspects," *American Biol. Teacher*, **25**, 409–411 (1963).

[9] Odum, E. P., *Ecology*, W. B. Saunders Co., Philadelphia, 1963.

[10] Odum, H. T., "Trophic Structure and Productivity of Silver Springs, Florida," *Ecol. Monogr.*, **27**, 55–112 (1957).

[11] Odum, H. T., R. J. Beyers, and Neal E. Armstrong, "Consequences of Small Storage Capacity in Nannoplankton Pertinent to Measurement of Primary Production in Tropical Waters," *J. Mar. Res.*, **21**(3), 191–198 (1963).

[12] Odum, H. T., A. Lugo, and L. Burns, "Metabolism of Forest Floor Microcosms," in *A Tropical Rain Forest*, H. T. Odum and R. F. Pigeon (eds.), Div. Techn. Information and Education, U.S. Atomic Energy Commission, Oak Ridge, Tenn., 1970.

[13] Redfield, A. C., *On the Proportions of Organic Derivatives in Sea Water and Their Relation to the Composition of Plankton*, James Johnstone Memorial Edition, University Press of Liverpool, 1934, pp. 176–192.

[14] Sverdrup, H. U., W. Johnson, and R. H. Fleming, *The Oceans*, Prentice-Hall, New York, 1942.

[15] Guyol, N. B., *Energy Resources of the World*, Dept. of State, Publication 3428, U.S. Government Printing Office, Washington, D.C., 1949.

[16] Wolman, A., "The Metabolism of Cities," *Sci. Am.*, **213**, 179–190 (1965).

[17] Woodwell, G. M., and A. H. Sparrow, "Effect of Ionizing Radiation on Ecological Systems," *Ecological Effects of Nuclear War*, ed. G. Woodwell, Brookhaven National Laboratory BNL 917 (C–43), 1965, pp. 20–38.

2

WHAT POWER IS

In human affairs the word *power* often describes the effectiveness of action or the capability of action. Great military power implies large military bodies involving many men and machines and exerting a directing force over large areas. Great political power suggests command of large numbers of votes and a wide influence on governmental systems and on the actions of many persons. Great economic power implies control of large volumes of money and of influences that can be bought with volume spending. Almost everyone understands in a qualitative way what power means in the affairs of man, but few equate concepts of general power with scientific thinking.

In the world of science and engineering, power is defined precisely in terms of measurable units as *the rate of flow of useful energy.* Scientific definitions of power are quantitative, and energy can be measured in such units as the calorie, the British thermal unit, and the erg. The flow of energy (power) is measured in time units such as *calories per day* or *horsepower* (1 hp = 10.688 kcal/min; 1 kcal (kilogram-calorie) equals 1000 calories; 1 BTU = 0.252 kcal).

The ability of a machine to accomplish a function is determined by its power rating, with a large number indicating a large role. For example, the functions provided by an air conditioner, a heating system, or a locomotive are described by their power deliveries. Although nearly everyone is familiar with power ratings of household appliances and automobiles, our educational system has rarely emphasized that the affairs of man also have quantitative power ratings and that the important issues of man's existence and survival are as fully regulated by the laws of energetics as are the machines. It is possible to put calories-per-day values on human institutions, on the flows of energy in cities, on the power requirements and delivery of activities of nations, or on the relative influences exerted by man and his environmental systems.

Most people think that man has progressed in the modern industrial era because his knowledge and ingenuity have no limits—a dangerous partial truth. All progress is due to special power subsidies, and progress evaporates whenever and wherever they are removed. Knowledge and ingenuity are the means for applying power subsidies when they are available, and the development and retention of knowledge are also dependent on power delivery.

Energetics, as applied to physical systems and machinery, is usually first presented in late grammar school. Because all phenomena of the real world, and not just machines, operate according to some basic energetic principles, it is at this educational level that the hard facts of power and its regulation of the larger environmental systems and human affairs need to be stressed. For lack of such education the public and its leaders are sometimes misled about their future.

First, let us consider the hard facts about energy flow that apply to all the phenomena on earth. Which energy laws are essential to the critical issue of man's continuing survival on this planet and his problems of war, peace, government, money, and food?

PRINCIPLE OF ENERGY DEGRADATION.

The most important hard reality about energy flow is the principle of energy degradation, illustrated by the energy flow diagrams in Figure 2–1. In Figure 2–1(a) water in the mountains contains potential energy which drives flow and restorage in the tank. For any process some of the available potential energy must be dispersed along the way as unusable heat. In other words, the potential energy of water in the mountains can be connected to some process if the arrangement provides for part of this potential energy to go to waste in the form of useless dispersed heat. In textbooks on energetics this principle is called the second law of thermodynamics. A general way of showing these processes is drawn in Figure 2–1(b). Energy waste dispersal is shown with sink symbols. In terms of further use the energy has gone "down the drain" as heat.

Heat is the state of energy consisting of the random motions and vibrations of the molecules. An object is hot when its molecules have much vibration and motion. These motions tend to spread from a hot body to a cooler one by random wandering and bumping that transfers and dis-

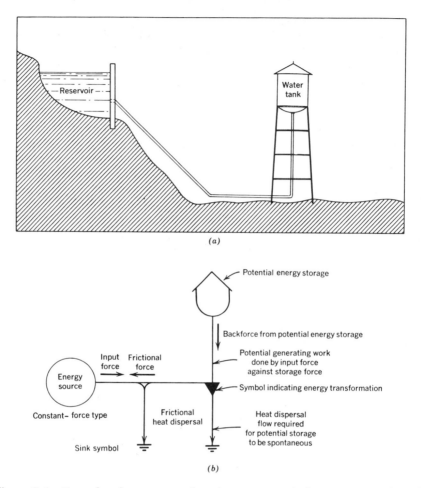

Figure 2–1 Example of an energy flow (power circuit) from a source through pathways that have frictional backforces to a point where there is a restorage of the energy, doing work against potential generating backforces. (*a*) Schematic picture of water system from the mountains with energy restored; (*b*) energy network diagram using symbols defined in Figure 2–4.

perses the energies of these motions. It is this random dispersal that ultimately is responsible for spontaneous energy transformations. The wandering pulls other processes coupled to it. Energy can flow only if some of it is dispersed by heat flow into the "sink." Without such energy dispersal, the downstream energy storages would push back with as much force as the upstream source. If we try to arrange an energy flow to operate so that the heat drain is eliminated, the process simply will not work. The heat drain is sometimes called the energy tax necessary for operation.

The energy that goes to the sink is not actually destroyed as energy but ends up as the speeding, bouncing, shaking, velocities of billions of molecules, each going in a different direction. This motion is disorganized, however, and cannot be harnessed except when one object has more of this energy than another next to it. If the energy that flows down the tax drain (Fig. 2–1(b)) is evenly dispersed in the environment as random heat energy, it can be said to be lost as potential energy, although technically it still exists as molecular motion.

The first law of energetics states that energy in processes not involving appreciable conversion of energy and matter is neither created nor destroyed. Thus, in Figure 2–1 all the calories flowing in from the potential energy source on the left must be accounted for in the storage and two outflows on the right. However, *potential energy*, which is the energy available to carry out additional processes and to account for more phenomena, is lost. It is degraded from a form of energy capable of driving phenomena into a form that is not capable of doing so.

Thus we may restate the important principle of energy degradation as follows: *in any real process useful potential energy becomes lost.*

THE ONE-WAY DIRECTION OF POWER DELIVERY

Because of the power drain required for a process actually to work, any operation involves a one-way processing of energy as the potential energy is removed from availability into dispersed unavailable form. Thus any procedure is unidirectional, and use and reuse of potential energy are not possible in the processes on earth. The unidirectional nature of energy processes is reflected in such common expressions as "One can't get something for nothing" and "Perpetual motion is not possible."[1]

WHAT FRACTION OF THE POWER GOES DOWN THE HEAT SINK

What fraction of the power flow must go down the compulsory tax drain? Is there any law that controls the quantity of this loss? Let us consider the arrangements of weights in Figure 2–2. In Figure 2–2(a) the weights

[1] Actually, perpetual motion in space, where there may be no friction, is not an energy transfer and is possible as long as other processes do not become involved. In earlier controversies, however, perpetual motion on earth implied the perpetual reuse, storage, and restorage of energy without inside input of new potential energy.

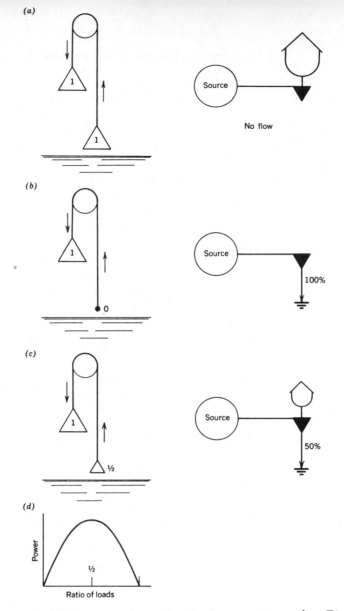

Figure 2–2 The fall of one weight supplies the force to raise another. Diagrams of energy transformations with three different conditions of loading of input and output weights. Energy flow diagrams are shown for each case. (*a*) Reversible stall; backforce and input force equal; no flow, no power delivery to new storage, the reversible case. If it moved it would be 100% efficient. (*b*) Free drop; no backforce; maximum flow rate, no power delivery to new storage, no provision for delivering load-lifting work. (*c*) Maximum power storage rate; backforce loaded to be half of input force; moderate flow rate, maximum power delivery to new storage, half of energy dispersed to heat sink. (*d*) Graph of power and load.

are arranged so that if the weight on the left falls, it would raise the same weight on the right. The falling of a weight represents a flow of power from the potential energy of its high position and is measured as the acceleration of gravity times the mass of the weight times the velocity of the weight. The raising of a weight is a flow of power into storage as the weight is lifted to a high position. With the weights equal, the system would, if it operated, convert all its potential energy inflow into useful, reusable, stored potential energy. However, for such an arrangement to result in motion and energy transfer is contrary to the principle of energy degradation. Common sense also tells us that the two weights, being balanced, remain still and nothing happens. Situations with balanced forces of opposing energy storages are said to be *"reversible."*

In the middle diagram (Fig. 2–2(*b*)) the weight on the right is zero. Now the weight on the left can drop without drag from any weight on the right. The potential energy stored in the weight flows rapidly out of its storage state into the kinetic energy of the falling weight. When the weight stops at the bottom, all the concentrated energy of the falling object goes into heating the object and the surface it hit. Very shortly that heat is dispersed evenly in the environment in the invisible motions of the molecules. Since there was no weight on the right-hand side, all the potential energy went into the heat drain and none into the useful storage process of elevating a weight. The example in Figure 2–2(*b*) is thus very fast, has a very large energy drain, and has no useful power output towards raising a weight.

In an intermediate arrangement (Fig. 2–2(*c*)) the weight on the left is about twice that on the right. In this example the weight on the left does fall, half of the energy being stored as available and useful potential energy for further processes, and half of the energy going into the heat drain and becoming unavailable for further useful processing. The arrangement in Figure 2–2(*c*) is the optimum one for processing the greatest amount of useful power at the fastest rate. If the loading weight is greater than one-half, the process of storage is much slower; any loading weight that is less wastes too much in the heat drain.

DARWIN-LOTKA ENERGY LAW

Thus, whenever it is necessary to transform and restore the greatest amount of energy at the fastest possible rate, 50 percent of it must go into the drain [14]. Nature and man both have energy storages as part of their operations and when power storage is important, it is maximized by adjusting loads, as demonstrated in Figure 2–2(*b*). In the last century

Darwin popularized the concept of *natural selection,* and early in this century Lotka [11] indicated that the maximization of power for useful purposes was the criterion for natural selection. Darwin's evolutionary law thus developed into a general energy law.

STORING WORK, PROCESSING WORK, AND ACCELERATING WORK

When power from a potential energy source is flowing through and driving a useful process, we are accustomed to describe it as "work." Thus work is done when weights are lifted, when objects are arranged, or when anything useful happens.[2]

It is important to distinguish among three kinds of work. The first, *storing work,* has already been illustrated in Figure 2–2. Maximum power flow in storing energy at maximum speed requires a 50 percent drain. Such power involves the force of one potential energy source directed against the back force from another potential energy storage. The energy cost of the process can be measured either by the energy stored or by the total power flow, which is twice the storage rate when the process operates in optimal fashion.

In the second kind of work, *processing work,* power flows from the potential energy source and passes through the system, arranging matter but effecting no storage and no final acceleration. All the energy eventually goes into the heat drain, but first useful ends are accomplished. For example, if we arrange books in a library or trucks in a parking lot, power is delivered into temporary motions which are then stopped, with energy being transformed into heat by the various frictions of the disordering and stopping motions. Arrangement of structure is a principal task in preserving or establishing any kind of order, whether it involves the maintenance of a living being or the operation of an industry. Thus work done against frictional forces to accomplish necessary and useful ends without storage leads to 100 percent drain. The energy cost of the process is not measured in the energy stored, for none is stored; but it may be

---

[2] The scientific, quantitative definition of work characterizes it as the product of a force times the distance over which that force operates. Any process involves a force and an opposing force, and any process involves work done in operating the forces for a distance. The force that is used to overcome resisting back forces flows from a potential energy source and is the means for delivering power. The opposing forces may be from other potential energy sources, from the friction that opposes any velocity, or from the inertial force that opposes any acceleration in the absence of other forces.

measured by the power delivered to accomplish the work. Processes of this kind represent much of the work of human beings, for example, the typing and filing of a secretary or the handiwork of a craftsman. In industrial planning it is customary to measure work of this type by the time required to do the job when efficient procedures are used. To convert data on work time into power figures, we may multiply time by the energy flow through the human being doing the work for a full day, for this flow includes the actual work plus the necessary maintenance of the human being's biological systems that allows the work to be done.[3]

A third kind of action processes the potential energy from its storage source into accelerating objects. Thus we may throw a ball and pass energy from storage into the kinetic energy of the ball. If the ball is thrown in a place where there is little or no friction, as in outer space, 100 percent of the energy may go into kinetic form. At first glance this conversion may seem an exception to the principle of energy degradation because no heat is dispersed into a drain. But this class of energy flow is only a relative energy transfer and not a true energy transformation. Einstein in 1902 indicated that a process of acceleration such as throwing a ball on one planet would look like deceleration when viewed from a second planet traveling by in a direction opposite that in which the ball was thrown. Any process whose direction depends on the place from which we are looking has, when viewed from the universe, no unique direction. Thus energy flows involving accelerations and work against inertial forces are not energetic transformations in the same sense that those requiring loss of potential energy are.

The action in acceleration can be considered work only in the sense of work done against an inertial force—but inertial force is defined, in turn, as a back force against acceleration. This circular reasoning started by Newton made it possible to state that a force always has an opposing force. Work done in acceleration requires no heat-sink dispersal and the energy stored in kinetic form is still available to drive other kinds of work. Therefore, it does not fall within the second principle of energetics which applies to processes with real power flows that exist regardless of the position from which we observe them. A pendulum operating in a vacuum is an example of energy changing its relative form from potential to kinetic without involving the second principle of energetics.[4]

[3] The power flow through an organism is approximated by measuring the oxygen consumed in the cold fire of the body's metabolism. There are about 4-5 k-calories processed per gram of oxygen consumed or per gram (dry weight) of food metabolized.

[4] In electric transformers energy is transferred from one electric coil to another by

Thus work done against inertial forces is a special exception to the energy degradation principle. When we accelerate matter, it is not like the true power delivery for useful processes requiring heat drain. When matter is accelerated, potential energy is changed from a resting form to one with a velocity relative to the observer. Thus kinetic energy and the energy in the magnetic fields about flowing electric charges are really interchangeable forms of potential energy. However, if we stop the motion with frictional force, energetic degradation results. Hence the many accelerations and frictional decelerations involved in most operations constitute processing work. In a real way, frictional forces are necessary to the processes of work that support man's order. If there were no friction, the order would be one of endless pendulum motions.

ENERGETIC DETERMINISM AND CAUSAL FORCE

In the affairs of forests, seas, cities, and human beings, the potential energy sources that are available flow through each process, doing and driving useful work of one of the types mentioned. The availability of power sources determines the amount of work activity that can exist, and control of these power flows determines the power in man's affairs and in his relative influence on nature. That any and every process and activity on earth is an energy manifestation measurable in energy units is a fact of existence. We may say that phenomena on earth are energetically determined.

The common use of the cause concept[5] is another manifestation of energetic determinism. As we have already stated, every power delivery from a potential energy source involves the expression of a force (X) against some opposing force. Power is the product of the force (X) and flow (J); see Equation 2–1. The amount of force can be scientifically defined by dividing the power (JX) delivered by the velocity of motion (J).

the surges of electric current which cause temporary magnetic fields created by one coil to induce a current flow in the other coil. By having coils of different properties, the voltage of the electrical power may be changed as desired. When side effects are eliminated, such transformers pass power approaching 100 percent efficiency. A person may wonder whether such energy transfers are exceptions to the second principle of energetics. Transfer of electric energy by transformer is yet another kind of acceleration of matter (charged electrons in wires), once again a relative process. As viewed from one position, energy changes its apparent form from electric potential to magnetic field energy. Viewed from a different position which is moving in the opposite direction, the transformation appears to be from magnetic field energy to electric potential energy. The energy transformation is only a relative one.

[5] See Reference 10 for a review of the concept.

Processes important to man are unidirectional because of the accompanying energetic degradation that makes them irreversible. We may designate the input force (X) as the causal force because it goes in the same direction as the flows.[6] Flows (J) induce friction but if there were no frictional heat dispersal the flow would not go spontaneously.

$$(\text{Power}) = (\text{Flow})(\text{Force})$$
$$P \quad = \quad J \quad\quad X \tag{2-1}$$

Frictional forces develop to balance input force and exist only against a flow. Since force (X) is proportional to the power delivery (JX), most of the everyday ideas of causal force are also statements of the ability of the energy source to deliver power and are adequately measured by energetic data. Units either of force (X) or of power (JX) can be used to measure causal action.

When we apply the concept of force to social groups, movements, and influences, we must consider program sequences and groups of associated physical forces involving power flows through a population. Whether we examine a simple process such as the fall of an apple or a complex cluster of millions of related processes such as a military campaign, we can measure the causal actions as power delivered from a power source through the action of forces. For the complex world of environmental systems and man, it is usually easier to measure the energies in fuels involved than to determine all the component and transient physical forces exerted.

The existing power budget in an unchanging system does not necessarily measure the power available to focus on a new causal action, although powerful systems are often capable of flexibility in the use of their flows or in tapping new sources as needed. Power to do new things depends on new power development or shift of power.

In complex social systems the patterns of cause are invisible until we draw a network of energy flow and accompanying forces. A vector is an arrow that expresses the direction and magnitude of force. Vector diagrams are often used to portray complex systems. Blalock [3] constructs

[6] In the past, some physics textbooks, in trying to emphasize the concept of balance of forces, made the point that there is no such thing as cause or effect. This statement is true only for problems of static force balance and accelerations not involving heat-dispersing work processes and energy degradation. Such teachings may have made sociologists, economists, and men in public affairs who had taken physics and science courses and learned about quantitative means for defining and measuring power limitations compartmentalize this knowledge and fail to use the concept of cause in the larger systems of nature and man where it is a reality.

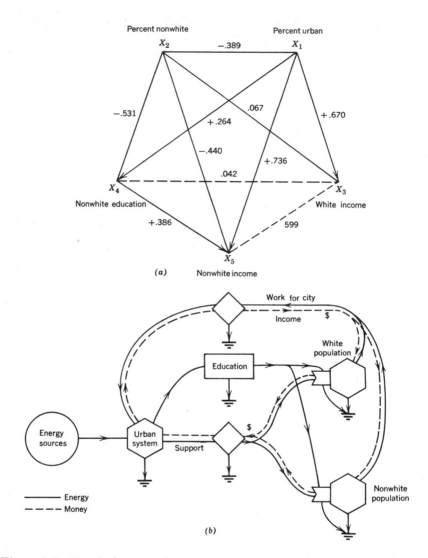

Figure 2–3 Causal diagrams for some population-education-income relationships. (*a*) Blalock's [3] correlations and resulting vector model of causal force to account for the correlations. (*b*) Energy-currency model for the same system as in (*a*). Force and energy pathways exist where there are positive correlations in (*a*). See explanation of symbols in Figure 2–4. Professor Blalock suggests the use of path coefficients also.

causal networks for human systems by drawing vector patterns between measurable entities. In statistics the tendency of one property to go with another is calculated by means of a formula called a correlation coefficient. Correlation coefficients are used by investigators to measure the strength of the vectors defining the causal network. Figure 2–3(a) shows one of these force diagrams, and Figure 2–3(b) shows possible pathways of these influences in energy flow.

Although these studies have not been stated in units of ordinary force or energy, the vectors do show common characteristics of these two ways of expressing pathways of causal force.

ENERGY NETWORK DIAGRAMS

For study and communication we now introduce the energy network diagram as our language to express energy flows and forces that represent the parts and relations of all systems. The lines represent the pathways of power flow and action of forces. The symbols are given in Figure 2–4. For those interested in the equivalent statements in traditional mathematics, explanations are given in the Appendix. Symbols of the energy circuit language have already been used in Figures 2–1 and 2–3. Flows of energy move from the sources, usually drawn on the left along the pathway lines to the right. We refer to these flows as moving from upstream to downstream when they move from source towards sink. In energy circuit language, the simple arrow is used to indicate the direction of the driving force. If there are driving forces pushing from both ends, no arrow is used. If backforces push but backflow is prevented, the pathway is a valve. A one-way valve symbol has the arrow symbol with a cross bar as in Figure 2–4(m). A square box is reserved for any unit for which there is not a special symbol; the unit is identified by a description inside the box.

ENERGY CIRCUIT OF A HUMAN HABITATION

In Figure 2–5 the energy network language illustrates and synthesizes the functions and relationships of a familiar system, the human house, which has been described as an ecological system by Ordish [15]. Some of the flows in a human house, such as the heating system, are nonliving processes. There are also living components such as populations of people and of mice. The diagram shows force relationships and relative magnitudes of flows and storages. Note the work done by humans in controlling and making the other flows possible.

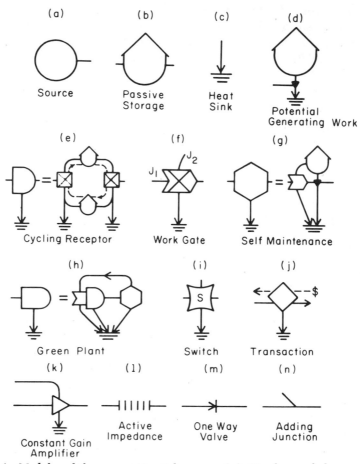

Figure 2–4 Modules of the energy circuit language. (*a*) Circular symbol represents a source of energy such as the sun, fossil fuel, or the water from a reservoir. A full description of this source would require supplementary description indicating if the source were constant force, constant flux, or programmed in a particular sequence with, for example, a square wave or sine wave. (*b*) Passive storage symbol showing location in a system for passive storage such as moving potatoes into a grocery store or fuel into a tank. No new potential energy is generated and some work must be done in the process of moving the potential energy in and out of the storage by some other unit. (*c*) Heat sink required according to the second law of thermodynamics for all processes that are real and spontaneous. All processes deliver some potential energy into heat. Heat is the random wandering of molecules that have kinetic energy and it is this wandering from a less probable to a more probable state that pulls and drives real processes connected to such flows. (*d*) Combination of (*b*) and (*c*) that represents the storage of new potential energy against some storage force, and such work requires a dispersal of some potential energy into heat to be

spontaneous. If this storing is done at the maximum possible rate, then 50 percent must be delivered into the heat sink. (*e*) The bullet-shaped symbol represents the reception of pure wave energy such as sound, light, and water waves. In this module energy interacts with some cycling material producing an energy-activated state, which then returns to its deactivated state passing energy on to the next step in a chain of processes. The kinetics of this module was first discovered in a reaction of an enzyme with its substrate and is called a Michaelis-Menton reaction. (*f*) Work gate module at which a flow of energy (J_2) makes possible another flow of energy (J_1). This action may be as simple as a person turning a valve, or it may be the interaction of limiting fertilizer in photosynthesis. (*g*) Hexagonal symbol represents the combination of (*d*) and (*f*) by which potential energy stored in one or more sites in a subsystem is fed back to do work on the successful processing and work of that unit. In its simple form this module is sometimes said to be autocatalytic. Its growth when graphed has a sigmoid pattern. (*h*) This symbol is a combination of (*g*) and (*e*). Energy captured by a cycling receptor unit is passed to self-maintaining unit that also keeps the cycling receptor machinery working, and returns necessary materials to it. The green plant is an example. (*i*) This symbol is used for flows, which have only on and off states controlling other flows by switching actions. There are many possible switching actions as classified in discussions of digital logic. Some are simple on and off; others are on when two or more energy flows are simultaneously on; some are on when connecting energy flows are off, and so forth. Many actions of complex organisms and man are on and off switching actions such as voting, reproduction, and starting the car. (*j*) This symbol is used for systems that have money cycles as well as energy flows. Money flows in the opposite direction to the flow of energy and the concept of price which operates among human bargains adjusts one flow to be in proportion to the other. Thus a man purchasing groceries at a store receives groceries in one direction while paying money in the opposite direction. The heat losses of these transactions are small since the work involved is small. If there are complex structures regulating the transactions the costs of the coupling may be greater. (*k*) As in (*f*) the work of one flow makes possible the work of a second, but in this module (*k*) the amount of energy supplied from the upper flow is that necessary to increase the force expressed by the system by a constant factor, which is called the gain. For example, a species reproducing with 10 offspring has a gain of 10 so long as the energy supplies are more than adequate to maintain this rate of increase. (*l*) This symbol represents the characteristic of many systems to develop a backforce against any input driving force so long as that force is increased. At the same time some energy storage is arranged so that when the forcing impetus ceases, the energy unit delivers a forward flow (from some storage or other source) in proportion to the earlier integrated impetus. Many organisms, human behavior programs, and conservative institutional programs have stubbornness of this form. In electrical systems such units allow time lags and oscillations to be designed or eliminated. (*m*) Allows flow to pass in one direction only even though there may be backforce from downstream storage. Symbols (*d-k*) also have the one-way property because of energy losses and interactions with second flows. If there are no backforces from downstream storages an ordinary barb is used without the bar. (*n*) Intersection of two flows of similar energy type, capable of adding. The behavior of this symbol may be contrasted with that of the work gate where the interaction is multiplicative. Each of these modules has characteristic equations which may be evaluated in real examples (see Appendix).

SUMMARY OF BASIC ENERGY FLOW DIAGRAM

The following are brief statements of some energy principles and the means for drawing them as a network using the symbols of Figure 2–4 and the example in Figure 2–5.

Everything and anything that takes place on earth involves a flow of potential energy from sources (*a*) into dispersed heat through pathways driven by directed forces which originate from energy storages (*b*). The

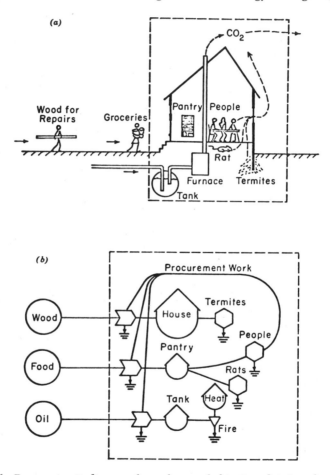

Figure 2–5 Power circuit diagram for a human habitation showing three outside energy sources, work by people, and steady-state maintenance of population and the structures of furnace and house. (*a*) Schematic diagram; (*b*) energy diagram; (*c*) flow of carbon.

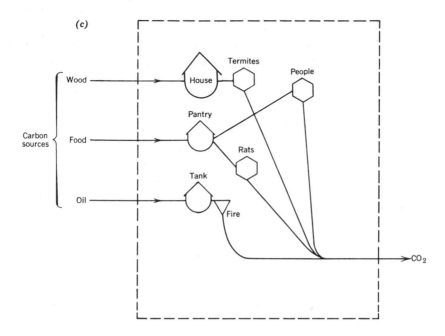

essence of cause, Newtonian physics, and the laws of energetics are irrefutable on these principles, but only recently have these laws that were developed for simple physical systems been applied to nature's complex ecological system or to the even more complex system of man's civilized activities in the biosphere. The flows of energy through complex food chains and cultural economic systems follow the basic laws, and we may use these quantitative relationships if we realize that the flows in the macroscopic world are primarily parallel flows of populations of large and small parts, including populations of molecules, cells, organisms, people, occupational groups, and other associations of active components.

The energy network language given in Figure 2–4 represents the networks of energy flows in a useful way and includes the requirements and limitations of energetic law. Here a pathway of potential energy flow is represented by a line, and if it is inherently unidirectional, incapable of reversal, it is marked with an arrow symbol (m). The potential energy storages are marked with the tank symbol (b and d) which indicates a source of causal force along the pathway lines. The force is in proportion to the storage whenever the storage function is one of stacking up units of similar calorie content. Most ecological and civilization systems have such storage systems.

We uphold the second energetic principle by drawing the network so that any process has some potential energy diverted into the dispersed random motion of molecules (heat). The downward flow into the heat sink is symbolized by an arrow directed into the ground. The first law of energetics is upheld by having all inflows balance outflows into storages, into the heat ground, or into exports. The many kinds of work done against frictional forces are illustrated by arrows that flow from the potential energy storage tanks into the heat sink. Whenever a work process is necessary for a second flow, facilitating the other flow by the work done on it, the arrow crosses the second flow. For example, the work of people on a farm facilitates the flow of light energy into food storage, although the energy of the workers' food which sustains them during the period of work is converted into heat by their metabolism (Fig. 2–4(f)). A work flow that facilitates a second flow in proportion to its necessary activity is mathematically a product function and thus is indicated by a box containing an X. Such a box indicates a controlling role of the flow in short supply. Such pathways are called limiting factors and the work can be thought of as a control valve gate on the other flow. Many of these control flows derive their driving potential energies from a point downstream and thus are multiplicative positive feedbacks (Fig. 2–4(g)). Whenever an energy flow is transformed and restored into a tank (Fig. 2–4(d)) at the optimum rate for energy storage, 50 percent of it must go into the heat drain. If pathways of a system must pass over hills or gaps where the flow must be pumped, there must be temporary additions of energy. These pathways are called energy barriers. An energy carrier that crosses the barrier must increase in potential energy temporarily and hence must lose 50 percent to the heat sink. An ordinary arrow symbol barb is placed on the line if the pathway receives no backforces from downstream storages and the only backforces are frictional. Other symbols are used in later chapters.

FLUX IN PROPORTION TO THE POPULATIONS OF FORCES

Each line of the energy network diagram represents a population of forces acting against frictional forces to cause a flow of energy either alone (as in heat flows, wave energy flows, or light) or associated with a flow of materials such as minerals or organic matter. For the energy diagrams of population phenomena given in this book, the important causal relationship that exists when driving forces balance friction is stated as

Flow (J) is proportional to population of active forces (N)

or

$$J \propto N \qquad (2\text{-}2)$$

where the forces are usually from populations of events working together or in succession. Equation 2-2 states that the rate of flow of such stored energy packages as minerals, money, or work, is in proportion to the number of forces acting as well as to their individual magnitude. For each separate component force the flow is in proportion to the magnitude of the driving force minus any backforces from downstream storages. For example, the total body metabolism, money expenditures, or nitrogen excretion of a population of people is in proportion to the number of people involved in the circuit.

For a circuit of particular conducting tendency, the proportionality in Equation 2-1 can be measured by L, the constant characterizing the friction in the pathway, sometimes called a conductivity:

$$J = LN \qquad (2\text{-}3)$$

The reciprocal of L is R, the resistance (R)

$$R = \frac{1}{L} \qquad (2\text{-}4)$$

An example from electrical systems is Ohm's law, which states that the flow of electrical current is proportional to the driving voltage (electrical force) according to the resistance (R) of the wire. Another example from groundwater geology states that the flow of water through the ground is proportional to the water level head according to the conductivity of the aquifer, which depends on its small channels through the earth. Chemical reactions depend on the population of reaction molecules along pathways that are dependent on resisting energy barriers and temperature. The identification of flows and forces in ecological and social systems is relatively new.

FACTOR CONTROL, LIMITING FACTORS, AND THE WORK GATE

Symbolized by the work gate symbol (Figs. 2-4(f), 2-6) are the many processes which have multiplicative actions when they interact. If a man turns a water valve, the water flows in proportion to the turning action as well as to the water pressure. The flow is a product of the two input factors. In this case, we often think of the more delicate of the two processes as the control. Work is done in the interaction and the process is controlled in the same way that a gate controls water in irrigation ditches. We can generalize the term "work gate" to refer to any multiplicative interaction of two flows and the forces they exert. As in all processes, potential energy is dispersed into the heat sink.

In chemical industry the reaction between two inflowing chemicals is proportional to the probability of the molecules encountering each other and this probability depends on their concentrations. The output of the reaction is a function of the product of the concentrations of the two chemicals at the reaction. This is an example of the work gate. As shown in Figure 2–6(a), varying either of the two reactants increases the output proportionately.

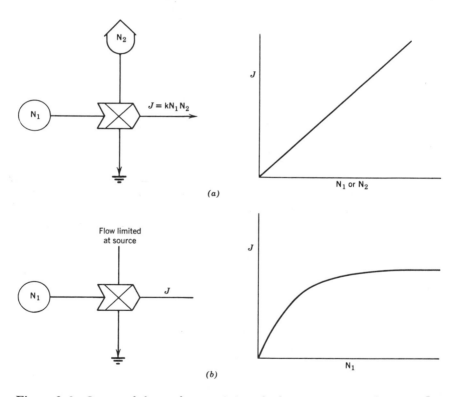

Figure 2–6 Output of the work gate. (a) Multiplicative response when two flows are maintaining local concentrations constant; (b) hyperbolic response when one of the flows is limited by its outside source or its recycling.

If, however, one of the inflowing reactants is limited at its source either from outside the system or from some restrictions in recycling within the system, then increases in the other concentration cause the output to increase but at a diminishing rate until the output is solely dependent on the limitation of the reactant in short supply. Shown in Figure 2–6(b) is the characteristic response of such curves which is called a rectangular hyperbola. Well established since the time of Liebig is the limiting factor concept by which the amount of some small limiting nutrient such as phosphorous determines the flow of some process such as agricultural production. Responses of plant production to nutrients follow the limiting factor hyperbola in Figure 2–6(b). Thus, the work gate symbol implies a limiting factor control action when the flows are limited.

Many of the limiting factors are also important energy sources to the processes although few people dealing with public resources have given them their appropriate calorie values. Many substances such as water and minerals have been regarded as free goods partly because their energy values were not realized and no dollar values were assigned until human work was involved.

VALUES FOR WATER AND OTHER CHEMICAL STUFFS

If a basic error in evaluation has been regarding natural sources as gifts rather than products from the ecosystem sector of our economy to be paid for, how do we evaluate the natural products such as water, air, oil, mineral deposits, and other chemical quantities? The science of biogeochemistry is concerned with estimating the quantities of materials cycling such as the water cycle, the nitrogen cycle, or DDT circulation. Data on worldwide and nationwide flows of some of these substances are already fairly substantial and in local areas it is possible to document accurately the flows of these materials. If we can obtain the potential energy value of each substance relative to the use that is to be made of it, we may multiply flow by kilocalories per gram to obtain the value of this material as a fuel to our usages.

We are not used to thinking of water, fertilizer, and air as fuels, but they are as much chemical reactants as oil. Any inflowing reactant to a process has its energy contribution to that reaction. There are chemical formulae for making these calculations if an approximate idea of the usage is known.

For example, rainwater picks up dissolved substances in going downstream in a watershed so that it may go from 1 ppm to 100 ppm by the time it reaches the delta. Then in passing from the delta to the open sea it picks up from 100 ppm to 35,000 ppm. The value of the water as a reactant for washing and dissolving stuffs and for many living functions has decreased in going downstream and is hastened by those who remove it for cleaning or for industrial purposes of this type. When calculations are made,[7] the potential energy value of the water as a fuel for washing processes is 10 times the value of the water in hydroelectric purposes (in areas of ordinary topography). Water can be more valuable as a chemical fuel than as a hydroelectric energy source.

The value of water can be dramatized by giving the energy value its approximate dollar equivalent in the overall economy (10,000 kcal/$). When this is done about 400 gallons of water are worth a dollar as industrial fuel. Contrast this with the usual cost figures for water, the cost of its processing, as 1000 to 5000 gallons per dollar.

Whereas such calculations give higher values for water, there are other usages of water with even higher energy values. The highest value may favor one usage in one location but a different usage elsewhere.

A municipality or an industry that uses public waters for some process with reactions, cleaning, or other usages that change its quality is using a public fuel, and public decision processes must require adequate payments and reimbursement to the former users, the sectors of the system who are now deprived, including the ecosystems as well as human sectors.

A reasonable plan of management of a system of man and nature requires that all flows be evaluated and paid for with return services and compatible inputs and outputs arranged. Similar kinds of calculations need to be made for all the chemical inputs from the natural and public sectors into our economies. The concept of free good must be eliminated.

DUAL ENERGY VALUE OF LIMITING FACTOR FLOWS

In the work gate interaction the inflowing material or service has two energy values. One is the potential energy flowing along its pathway which it brings to the work gate. For a small limiting factor such as

---

[7] Substituting in the expression for chemical potential energy we find 0.18 kcal/g of dissolved solids in going from rainwater to seawater. This is about 23 kcal per gallon (Odum [12]).

$$\Delta F = \Delta F_o + n\ RT\ \ln C_2/C_1$$

$$\Delta F = 0 + (\frac{1}{35}\ \text{mole})(1.99\ \text{cal/Deg. mole})\ (300)\ (\ln 1\ \text{ppm}/35{,}000\ \text{ppm})$$

Somewhat different values result depending on the kinds of molecules dissolved in the water as it is used in industrial and living activities.

Table 2–1   Comparison of Some Fuels and Consumers

System	Fuel Value per Volume (kcal/cm³)	Fuel Consumption Rate (kcal/day)
Green plant using sunlight (covering a square meter of ground)	6×10^{-16}	4,000
Waterfall (10 m high, 10^5 gallons per day) consuming energy of elevated water	2×10^{-5}	9,100
Human consuming food	4.5	3,000
Small automobile consuming gasoline	10	900,000

phosphorus for plant production this inflowing energy may be small. At the work gate, however, the limiting factor is the missing ingredient for a flow of much larger power of the plant production using such factors as light, water, and carbon dioxide. The limiting factor while it is limiting gains a large amplification factor causing much power flow to pass. The energy value of its amplification in the work gate is a second kind of energy value. To characterize the energy values of complex systems we need to evaluate both.

For example, Alkire [1] found 1.5 pounds of fish caught by atoll inhabitants per hour of effort. For about 125 kcal of food energy spent in fishing, a pathway yielded about 720 kcal, an amplification factor of about 6 times. The hourly energy flow value of this work was 125 kcal when not directed in the system reaction, but is 720 kcal after amplification in the work interaction.

POWER SOURCES

Potential energy available to produce causal forces and drive useful processes such as those supporting man and nature enter the realms of man from several sources. As shown in Table 2–1, some are concentrated and some are dilute. The coal and oil that were stored in limited quantities in the available crust of the earth are classified as concentrated types because they have large potential energies per volume (10 kcal/cm³; Table 2–1). Sunlight energy, although vast in quantity, flows in a dilute form and the amount per volume at any one time is only 6×10^{-16} kcal/cm³.[8] Water power is intermediate in concentration.

[8] Light energy per unit volume is related to light pressure and is a potential energy.

Nuclear energy is extremely concentrated because it involves the vast storages of energy in atomic structure, but we cannot arrive at the final value of its concentration to man's world until we subtract the energetic costs of maintaining systems for setting it free and protecting our system from radioactivity. It is difficult to predict what the net value of nuclear energy will be.

The energy of the sun, driving the atmospheric water engine by means of rain, maintains waters at high mountain elevations, thus providing water power potentials in many areas. Some of the products of nature involve concentration and conversion of light energy into food and forest wood which are highly concentrated energy sources. These concentrations are not produced rapidly, however, and thus cannot supply large power demands.

Much of the optimism in this century of rapid industrial progress has been based on the belief that energy sources were increasing along with the use of them. Many of the warnings that the opposite may be true have been ignored because each estimated date for the exhaustion of energy sources has been put off by new discoveries of oil and coal and by the development of atomic energy. There is a new fear of too much power.

The actual proof that atomic energy can replace fossil fuels (coal and oil) at an equally concentrated level of potential is difficult to obtain, even though some large atomic power plants are being built at apparently cheaper financial costs than those of fossil fuel competitors. The degree to which there are hidden subsidies from the main cultural fuel supply and from the ecosystems to the new and, at present, still minor nuclear sources is not readily measurable. If there were no energy-rich parent society to maintain a high degree of technology, could the many special features of atomic power plants and their human operations be supported? In other words, to what extent do the dollar costs of machinery and human resources used by the atomic power plants really represent the total costs?

CONSTANT-CURRENT AND CONSTANT-FORCE SOURCES

Energy flows may be further classified according to the properties with which they respond as we add a work load. Some sources are provided with a means for keeping the current constant. For example, an old-fashioned waterwheel dipping into a steady river flow may take varying degrees of energy from the river, but there is no change in flow rate. Another example is the steady fall of leaves in some tropical forests providing energy to the soil organisms. Usage does not increase the power. Sunlight is another example.

However, if a dam has a very large reservoir holding the water pressure constant, we can open various turbine tunnels to obtain the same pressure in each tunnel. As long as we do not draw enough water to change the water level, the force exerted will be constant. The energy source in this instance is a constant-force source. Small drains on large batteries are pushed with constant force.

Most sources of energy are classified as between the constant-current and constant-force types.

WHAT ARE SOME MAGNITUDES OF POWER DELIVERY?

Since we rarely think of some of the phenomena of the larger world systems in the same energy scale, it is useful to study comparisons such as those in Tables 2–1 and 2–2. In order to consider energies of large and small processes together, all power figures are given in the kilocalorie unit that is already familiar to most people from dietary uses in estimating the amount of potential energy in foods. Thus, most people know that something on the order of 3000 kcal[9] of potential energy is required each day for an adult person leading an average, active physical and mental life.

The daily energy levels reaching the earth and penetrating to the ground are in thousands of kilocalories per square meter and these energies drive the atmospheric engines, the ocean currents, the world's photosynthetic production and, ultimately, this energy is reradiated into space in all directions as invisible heat radiation (infrared).

The photosynthetic process concentrates the dilute solar energy achieving as much as 131 kcal/(m²)(day) in such productive land systems as rich forests and cultivated agriculture during their maximum growth periods. Estimates for the overall photosynthetic production for the whole earth, of which 71 percent is ocean, are much lower—about 6 kcal/(m²) (day) (Table 2–2).

These solar sources of energy for foods and fuels may be compared with the consumer systems. Made of complex organisms, humans, and

[9] This unit is the kilogram calorie which is 1000 times larger than the gram calorie, which is the heat that raises a cubic centimeter of water 1°C when the water is 15°C.

The calorie value in foods is often determined as the heat actually released when food is burned; this heat includes some that is not available as potential energy but comes from molecular state changes necessary in burning. Thus calorimetric data are not exactly equal to potential energies, but the error in using such data to measure potential energy in food materials is normally less than 10 percent. The potential energy of chemical storages available to processes for which pressure is held constant is called free energy. Fuel and food processings on the surface of the earth under a steady atmospheric pressure of air are of this type.

Table 2–2   Comparison of Power Flows of Some Systems

System	Power (kcal/m²/day)
Incoming energy (dilute type)	
Sunlight absorbed by biosphere	5110[a]
Sunlight reaching green plant level	3400[a]
Maximum conversion[b]	170
Organic production as in Figure 1–5	
Production of a rain forest	131[c]
World primary production	6[d]
World agriculture on 8.9% of the world's area[e]	0.26
World agriculture contribution to biosphere[e]	0.024
Production removed by farming[f]	3.6
Consumer systems	
Respiration of a rain forest	131[c]
A village of people without machines, 100 m²/person	30
Fossil-fuel consumption in the whole biosphere[g]	0.135
Total consumption of biosphere (production and fossil fuel)	6.1
Fossil-fuel consumption in the United States (per U.S. area)	3
Animal city, Texas oyster reef[h]	57
Fossil-fuel consumption in a large city[i]	4000

[a] Sellers [17]; 3.67×10^{18} kcal/day; 5.1×10^{14} m²/earth, 29% reflected [8].
[b] At high light intensities 10% of visible light is converted to organic matter; visible is 50% of total sunlight.
[c] Odum and Pigeon [13].
[d] Hutchinson [9]. This figure is defined as the sum of daytime net photosynthesis measured by metabolic substances released or absorbed by the medium of the plant cells (such as oxygen) plus the nighttime respiration. The actual gross photosynthetic process is much greater than can be measured in this way because there is fast internal cycling (making and using) in the day stimulated by photorespiration and by the impetus of storages in the various sites of reaction in the cells of the plants. Adding the night respiration is an objective procedure but one that apparently underestimates the amount of the daytime cycling. The figure given is much too small, therefore, as a measure of gross production.

In recent years maps of the world's primary production have been published using much lower values from radiocarbon uptake studies in bottles of seawater from the tropical oceans. These world averages are one-tenth of the above figure. In one respect most of these radiocarbon measurements were mostly erroneously calculated so that they represent neither the basic gross inflow of carbon into the cells (J in Fig. 11–1), nor the net accumulation ($J-kQ$ in Fig. 11–1), but something in-between. As measures of the food available to other food chains, they are useful, however, showing how little of the gross primary production gets outside of the plant cells over much of the earth. The actual net production is even less than these maps show.
[e] U.S. Department of Agriculture [18]. Production is first calculated per unit of

machines, high-quality mechanisms require high-quality food and fuel energies either from photosynthesis or from fossil fuels (4–10 kcal/cm^3; Table 2–1). Whereas the overall consumption rate of the biosphere may be near that of the primary production, much higher rates develop where consumers cluster their work in their characteristic way. Productive forests support much higher respiration rates than the average of the biosphere. Animal and human cities have very high power flows, requiring much air, water, the foods from large areas, and large fuel supplements from the fossil storages of oil and coal. Because of the high power demands of the complex consumer animals and machines (Table 2–1), there is a limit to the number which can be supported without serious imbalance of gaseous and mineral cycles and waste of high-grade energies.

Agriculture represents 8.9 percent of the world's photosynthetic surface. Although agricultural productions do approach the maximum yields of nature, their annual contributions to the earth's production is much less than the natural systems they replaced because crops are in full leaf only part of the year. As Hutchinson [9] showed, the removal of natural terrestrial production for agriculture has been a greater transient in the world carbon cycle and the world carbon-dioxide levels than the fossil-fuel contributions. Because of the higher net production of ecosystems on land than in much of the seas, removal of land from production is about five times more important than the same amount of water-covered surfaces.

A SURGE OF FOSSIL-FUEL AND POPULATION

As shown in Figure 2–7 there are accelerating rates of power consumption and with it population explosion, the impending crisis of our times. Not only do the industrialized countries expand their fuel demands and explorations, but the less developed countries import the system to catch up and expand. The oil industries accelerate oil exploration to keep up with the demand. Populations respond to cheap, high-grade energies with increased reproduction and survival which in turn accelerates demand. Many features of our culture and political life are presently geared to

---

agricultural land (4.55 × 10^{13} m^2) and then per area of the biosphere (5.1 × 10^{14} m^2).

f [40 kcal/(m^2) (day)] (Agricultural lands)/(area of earth).

g Revelle, Broecker, Craig, Keeling, and Smagorinsky [16].

h Copeland and Hoese [4].

i 0.025 persons/m^2; 1.6 × 10^5 kcal/(person)(day) (see Fig. 6–6).

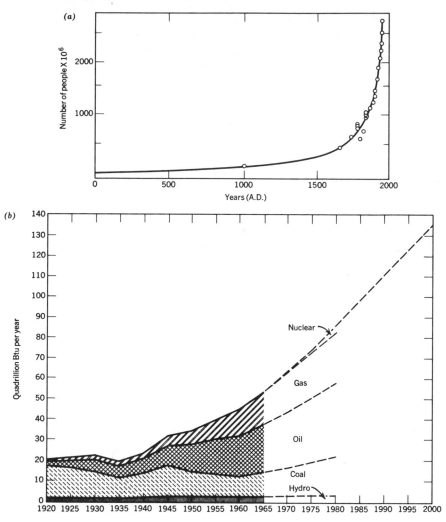

Figure 2–7   Exponential Malthusian growth. (*a*) World population (Von Foerster et al. [19]). (*b*) U.S. Energy Consumption [2].

this acceleration—for example, the oil depletion allowance which encourages oil development by tax exemption. Such circular stimulation is called positive feedback and mathematically produces exponential shaped growth curves as Malthus showed long ago.

Comparing the power levels in Figure 2–7 with those in Table 2–2, we find the industrial power budget in 1962 already approaching 8.1 percent of the world's organic budget of production and respiration.

UNDESIRABLE WORK, SHORT CIRCUITS

As power levels increase, problems with excess power are developing. A flow of potential energy is either stored or drives a work process. When storages are large, there is an immense concentration of ability to drive processes that will dissipate the energy accumulation. If no special protection for the storage energy has been arranged, opportunistic circuits may be connected and the energies discharged in disorder. Storage packages in nature require special protections, and most animals and plants divert energies into such protection, into special skin, bark, or individuals specialized to protect the group. In human systems the police perform one such function by protecting storages. The costs are the necessary part of having such a network.

A transformation of potential energy stalls if it is overloaded by efforts to accomplish too much potential generating work but, if underloaded, it tends to develop additional work. If we arrange for a flow to disperse the energy directly into the environment, the heat releases will become so concentrated that localized heat gradients will develop to do work while the energy is dissipating. The electric short circuit, the forest fire, and the waterfall are examples of potential energies released into heat with resulting eddies, structural damage, or work done on the environment.

A related example is found in human populations when foods are made available without work being required in payment. Suppose a food supply is sent free to another country, thus supporting a population without demanding a work loopback to the first system to make a harmonious economic complementary payment. The idle population does not have to engage in the regular processes for food output and hence tends to set up all kinds of unorganized activities. If no organized outlet for their contributions is arranged, unrest, mob actions, and social eddies can result, at least until a new closed-loop system evolves by selection. The giveaway of food in human affairs can be an energy short circuit.

COMPETITIVE EXCLUSION PRINCIPLE

The ecological principle of competitive exclusion [7] states that one of two populations of self-duplicating organisms, if permitted to expand with Malthusian growth without selective control, will drive the other out of the realm. The one with the slightly better reproductive performance is self-amplifying and the discrepancy between the two stocks increases. This is a well-established phenomenon in laboratory experimentation where other controls are removed. An example is given in Figure 2–8 where two grains were planted successively. This tendency for runaway competitive exclusion of one part of a network is a fearsome, ever-present danger against which all surviving systems must be protected by organizing influences.

CANCER AT MANY LEVELS

When a well-organized system is disrupted and its controls are destroyed, the parts may go into Malthusian competitive exclusion and in the process destroy the remnants of the system and itself. When this happens with cells in humans, we call it cancer.

The exact causes of cancer are difficult to identify primarily because so many influences are capable of disordering the normal control system. Radiations, chemicals, and senescence can interfere with the controls and release the dreadful competitive exclusive growth. Once started it gains control of the energy sources. The disturbing agents may be additive.

Our environmental crisis with its many stresses and disorders is producing cancer at many levels. Its chemicals are causing cell cancers and the disordering of the ecosystems is releasing well-controlled networks into widely exploding erratic growths of species, that are as cancerous to the ecosystem as are cancerous cells in the body.

The biggest cancer of them all is the human population itself. Somehow removed from its normal control network by modern medicine, and many other drastic operations using fossil fuel, some of the human populations have gone into competitive exclusions consuming resources, setting up subcompetitions in their cultures and races, and generally draining the greater body of the world towards its collapse. How do we restore the controls and cut off the fuel supplies to the runaway components? What right have oil companies to open Alaskan oil?

SUMMARY

The phenomena of the biosphere, including nature and man, can be measured and represented by the pathways of power which form systems that can be diagramed with energy flow diagrams. The economic, political, and social power flows are just as measurable as those of the simple physical and chemical world. Relative magnitudes of processes are comparable on a $kcal/(m^2)(day)$ basis. The basic energy laws of conservation, degradation, maximum power selection, proportionality of flux and forces, and solar budget provide hard facts of life for survival in the

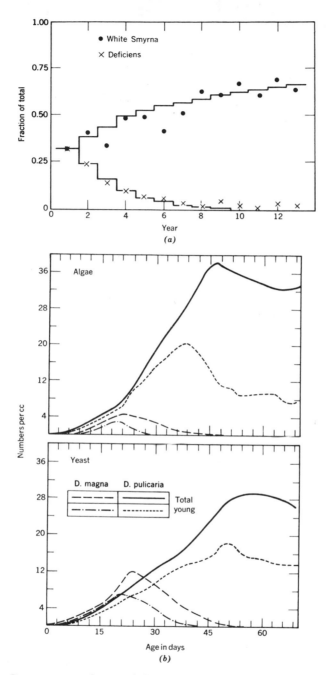

Figure 2–8 Competitive exclusion. (*a*) In grains successively replanted (DeWit [5]); (*b*) water fleas eating yeast and algae (Frank [6]).

biosphere. The demands of man and his machines for high energy of concentrated type contrast with the broad dilute field of incoming light energy from the sun. The surge of industrial and population growth is derived from fossil fuels now being consumed at accelerating rates (Fig. 2–6). The relative magnitudes of these new power flows, while still only a fraction of that of the whole biosphere, begin now to be sufficiently large to change and disturb the earlier checks and balances of the earth system wherever they are concentrated.

REFERENCES

[1] Alkire, W. H., "Lamotrek Atoll and Inter-Island Socioeconomic Ties," *Illinois Studies in Anthropology,* University of Illinois Press, 1965.

[2] Anonymous, *Energy R & D and National Progress,* Supt. of Documents, Washington, D.C., 1966.

[3] Blalock, H. M., *Causal Inferences in Experimental Research,* University of North Carolina Press, Chapel Hill, N.C., 1961.

[4] Copeland, B. J., and H. D. Hoese, "Growth and Mortality of the American Oyster *Crassostrea virginica* and High-Salinity Shallow Bays in Central Texas," *Publ. Inst. Mar. Sci.,* 11, 149–158 (1967).

[5] DeWit, C. T., "On Competition," *Versl. Landbouwk. Onderzoek,* No. 66.8 (1960), Wageningen, Netherlands.

[6] Frank, P. W., "Coactions in Laboratory Populations of Two Species of Daphnia," *Ecology,* 38, 510–519 (1957).

[7] Gause, G. F., *The Struggle for Existence,* Hafner, New York, 1934.

[8] Haar, T. H. Vonder, and V. E. Suomi, "Satellite Observations of the Earth's Radiation Budget," *Science,* 163, 667–668 (1969).

[9] Hutchinson, G. E., "The Biochemistry of the Terrestrial Atmosphere," in *The Earth as a Planet,* G. P. Kuiper (ed.), University of Chicago Press, Chicago, 1954, pp. 371–433.

[10] Jammer, M., *Concepts of Force,* Harper and Brothers, New York, 1957.

[11] Lotka, A. J., "Contribution to the Energetics of Evolution," *Proc. Natl. Acad. Sci.,* 8, 147–155 (1922).

[12] Odum, H. T., "Energy Values of Water Resources," *Proc. 19th Southern Water Resources and Pollution Conference,* Duke Univ., in press.

[13] Odum, H. T., and R. F. Pigeon (eds.), *A Tropical Rain Forest,* AEC Division Technical Information, Oak Ridge, Tenn., in press.

[14] Odum, H. T., and R. C. Pinkerton, "Time's Speed Regulator: the Optimum Efficiency for Maximum Power Output in Physical and Biological Systems," *Am. Scientist,* 43, 331–343 (1955).

[15] Ordish, G., *Living House,* Lippincott, Philadelphia, 1960.

[16] Revelle, R., W. Broecker, H. Craig, C. D. Keeling, and J. Smagorinsky, "Atmospheric Carbon Dioxide," in *Restoring the Quality of the Environment,* Report of President's Science Advisory Committee, Washington, D.C., 1965, pp. 111–133.

[17] Sellers, W. D., *Physical Climatology*, University of Chicago Press, Chicago, 1965.

[18] U.S. Department of Agriculture, *The World Food Budget 1970*, Foreign Agricultural Economic Report No. 19, 1964, 105 pp.

[19] Von Foerster, H., P. M. Mora, and L. W. Amiot, "Doomsday: Friday 13 Nov. A.D. 2026," *Science*, **132**, 1291–1295 (1960).

3

POWER IN ECOLOGICAL SYSTEMS

The study of such aspects of nature as forests, seas, and living reefs is the science of ecology. It is the study of ecological systems. In recent years the power flows in many ecological systems, including some of the forests, prairies, and savannas where man was originally contained as a minor component, have been examined and measurements have been made. Studies of the environmental energetics of the simpler networks of nature help us visualize, simplify, and perhaps see how to generalize about the more complex systems dominated by industrial man and his machines.

DEFINITIONS: THE ECOSYSTEM, SUBSYSTEMS

In Chapter 1 we found that the biosphere as a whole was a single system with gaseous, aqueous, and mineral circuits exchanging materials throughout. When we examine the system closely, looking at one part of the biosphere at a time, we can recognize spacial subdivisions of the overall system such as forests, ponds, and seas. Looking even closer, we can consider part of a forest, a pond, or a sea as a small system also—the bottom sediment, the productive zone receiving light, or the soil, for example. Thus the overall system can be viewed as subsystems within subsystems.

Therefore, for our studies of the structure and function of the macroscopic world, let us define systems by imposing arbitrary boundaries as shown in Figure 3–1. When the system is ecological, involving living components, it is called an ecosystem. As emphasized in later chapters, man, his machines, and his networks of communication and currency are part of the ecosystem and belong in energy diagrams of it. The systems

58

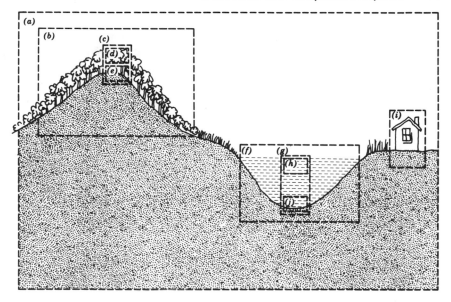

Figure 3–1 Forested mountain, fertile lake, and house with various ecosystems arbitrarily defined according to the purpose of study. (a) An inclusive system with the total landscape included; (b) a forested mountain; (c) a single prism of forest for intensive study; (d) a photosynthetic zone; (e) includes regenerative soil, litter, and animal zones; (f) a lake as a whole; (g) a vertical prism for representative calculations of some of the main parts of the lake; (h) the euphotic (lighted photosynthetic) zone subsystem; (i) the house further elaborated in Figure 2–5; and (j) mud and hypolimnetic subsystem (colder section of a density-stratified aquatic ecosystem).

of the world such as seas, cities, and savannas have structures and processes that blend into adjacent nature, often without discontinuities and only rarely with distinct boundaries. For the study of such phenomena we may precisely define ecosystems to include smaller or larger portions as pictured in Figure 3–1. Most small sections of nature studied as ecosystems are also parts of larger ecosystems and, in turn, contain smaller sections that can be studied as subsystems.

To understand ecosystems we must know two kinds of processes: (a) the rates of circulation within the boundary chosen, and (b) the rates of flows of energy and matter across the boundaries into and out of the system chosen. We describe the ecosystem environment when we analyze these boundary fluxes. When we deal with large inclusive ecosystems such as Figure 3–1(a), the boundary fluxes may be small compared to the internal cyclic fluxes; but in small ecosystems such as Figure 3–1(h), the boundary fluxes may predominate. To understand the mechanisms for the total performance of the overall system, we must have data on both the total system and main parts which we call compartments.

In each subsystem that we choose to study we may find circuits between compartments and circuits within compartments. Some circuits are species. In many patterns the subsystems carry miniaturized equivalents of the overall biospheric system. For example, the pulse of photosynthesis and oxygen production in such microcosms as lakes or small living laboratory terraria (Fig. 1–5) resemble those of the biosphere (Fig. 1–5). Thus there are similar processes and patterns of production, respiration, maintenance, and chlorophyll with space and time. Although the ecological systems (subsystems of the biosphere) differ considerably from each other, most have enough features in common to justify describing some basic patterns that occur frequently the world over.

Whereas ecosystems are divided into smaller ecosystems geometrically as in Figure 3–1, subsystems may also be separated for study in other ways when the details are of special interest. The subsystem of each species population can be isolated as is often done when man is considered without nature. Or, as illustrated in Figures 1–1 and 1–5, each mineral cycle (i.e., carbon, nitrogen, phosphorous, oxygen) can be separated for diagramming and study. We must remember that these are only partial systems. In man's study of himself he has often made his systems study with the economic consideration of currency only, thus omitting much of his system. Little wonder that his understanding and predictions of his economy have been faulty.

NETWORKS OF SPECIES

An ecological system is a network of food and mineral flows in which the major pathways are populations of animals, plants, and microorganisms, each specialized to live in a different way, doing a different job for the energy flows of the system. A species of living organism is a population that is insulated and kept separate by various means so that the pathways of its food and mineral flows are not entangled with those of other species. The separation of species is maintained by the various mechanisms that prevent interbreeding and mixing of genes, but the purpose of having many species is also the increased system effectiveness possible with specialization, division of labor, and more kinds of controls and regulatory circuits. A system with more species and hence increased organization may be compared to the radio that has more tubes. More circuits are designed to regulate and stabilize the overall function of the system. A network of species may in many systems involve several hundred different populations, although only a few of these are present in large numbers with conspicuous dominance. The dominant species are the power circuits. In conformance with the energy laws given in Chapter 2, we may diagram the network of power flows with one pathway junction for each species population.

Although the fishes and plankton of the sea seem vastly different as organisms from the birds and tree leaves of the land system, the energetic properties in the stage diagrams are quite similar. Comparing plankton in the sea with the leaves of the rain forest is a famous thesis of Bates [1] and Steeman-Nielsen [33].

In Figure 3–2 data and diagrams on the separate species and occupational circuits are shown for a marine estuary in Louisiana studied by R. Darnell [7, 8]. At this close proximity man begins to see so much beauty and magnificence in detail that he sometimes can no longer discern the overall pattern that explains the why, the functional design, and the great scheme of group influences guiding the separate species. There is little wonder that for many centuries people viewed the species nonfunctionally as a repatriation from Noah's ark, as God's handiwork to entertain man, or as an accidental result of genetic processes operating among islands. Separation of functions among populations increases efficiency by division of labor and provides specialists for controlling system functions. If species in nature and occupations among humans are both means for specialization, we can apply to the industrialized city some of the concepts used to understand the simpler ecological systems.

GROUPING INTO COMPARTMENTS OF SIMILAR FUNCTION

Unfortunately, data that show all the circuit relationships for the less common species are rarely available. In many studies, however, it has been possible to lump together functions or species that have properties in common or that can be measured as a group. Such grouping is called compartmentalization.

As we follow input energy from the source through a species population to a second population which eats the first and then to a third population, and so on, we encounter successive *stages* of energy use and storage. We may group together the populations that are in the first stage of energy processing, and those that are in the second stage. Be-

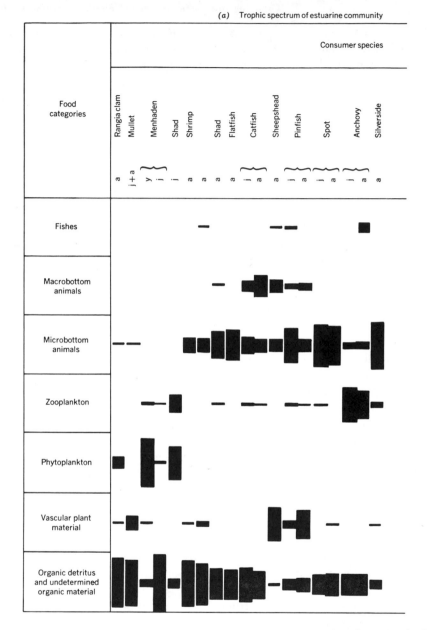

Figure 3–2 Energy flow diagrams for a marine estuary system based on the food chain matrix provided by Rezneat Darnell for Pontchartrain Estuary, Louisiana. (*a*) Data on food of species; (*b*) energy flow by species; (*c*) energy flow by habitat

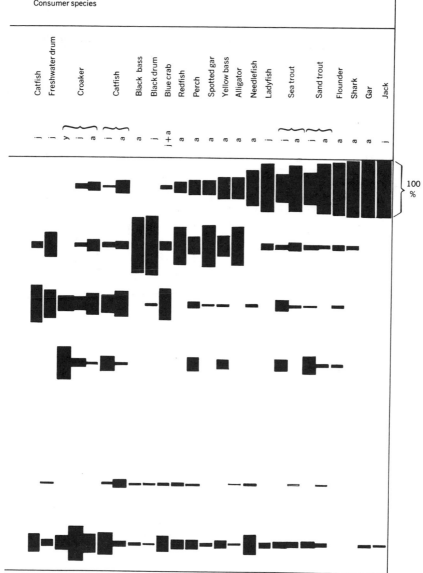

Consumer species

Catfish
Freshwater drum
Croaker
Catfish
Black bass
Black drum
Blue crab
Redfish
Perch
Spotted gar
Yellow bass
Alligator
Needlefish
Ladyfish
Sea trout
Sand trout
Flounder
Shark
Gar
Jack

100 %

compartments; (*d*) energy flow by trophic roles; (*e*) energy flow by trophic levels without organic source and pools; (*f*) *P* and *R* compartmentalization.

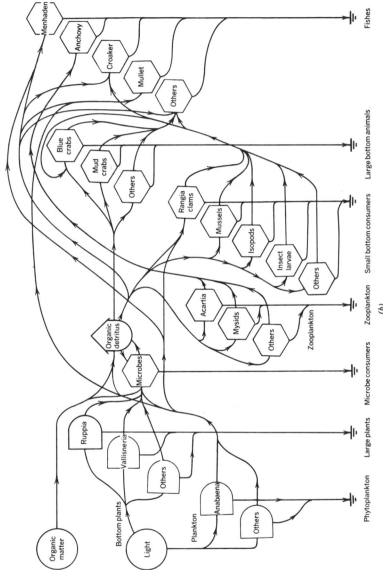

Organic matter

Bottom plants

Light

Plankton

Ruppia

Vallisneria

Others

Anabaena

Others

Microbes

Organic detritus

Acartia

Mysids

Others

Rangia clams

Mussels

Isopods

Insect larvae

Others

Blue crabs

Mud crabs

Others

Menhaden

Anchovy

Croaker

Mullet

Others

Phytoplankton   Large plants   Microbe consumers   Zooplankton   Small bottom consumers   Large bottom animals   Fishes

Zooplankton

(b)

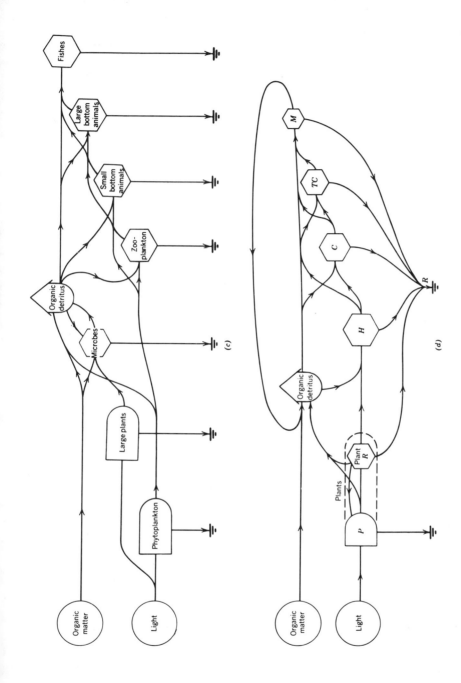

(c)

(d)

65

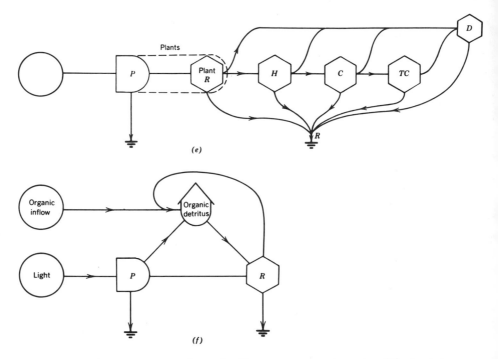

(e)

(f)

cause species are grouped, such diagrams present a simplified picture of the fractions of power that pass each stage and drain off in dispersed heat during the various work processes of the populations.

The system as a network of a few compartments can be drawn to reveal some basic energetic properties, such as the requirement that much of the food-fuel energy passing through each compartment be dispersed as heat, thus being unavailable for further use as a fuel. The compartment view has been very useful in computing broad features of systems such as rates of carbon flow, the rate of removal of some radioactive contaminants, accumulation of DDT, and the general carrying capacity of the earth for creatures of various classes such as herbivores, carnivores, and second-order carnivores. This grouping fits plankton systems of the sea as was evaluated in historically important examples by Clarke [5]. Thomas Park in his presidential address to the Ecological Society of America described this intermediate view as the broad-brush picture. Broad-brush views have also been used in economics, and in a later chapter an effort is made to relate the flow of money and the flow of materials and energy. It is enough to say now that the simultaneous flows of money and of potential energy and material go in opposite directions.

In Figure 3–2(b) a complex estuarine ecosystem studied by Darnell [7, 8] is diagrammed to show increasing compartmentalization. For simplicity, outflowing export losses of organic matter and the flows of inorganic minerals are all included in the pathways of respiration. In the diagrams these drop vertically into the heat sink. In Figure 3–2(b) some simplification is achieved by lumping the least common species in the miscellaneous category "Others."

In Figure 3–2(c) further compartmentalization has been arranged grouping populations with similar habitat localization (i.e., plankton or bottom). An alternative way of grouping into compartments is by general overall food role. Whereas it is well established that most species usually receive food contributions from organic pools, animals, plants, and microbes rather than from just one of these categories, it is still possible to classify a species by the food category that is largest in its diet. Figure 3–2(d) further simplifies the model of the ecosystem retaining the familiar categories of herbivore, carnivore, top carnivore, and miscellaneous, which includes microbial consumers and processors. This diagram is a fairly general one characteristic of many ecosystems, for it shows the importance of the organic detritus in all the food pathways.

An historically important idea in the period when ecologists were learning to compartmentalize and model is represented by the diagram in Figure 3–2(e) and shows an ecosystem divided into six divisions called trophic levels with each one step further removed from the source of energy and each having a drop of about 10 percent in potential energy flow remaining to the next level downstream. This concept from Lindeman can be implemented with data if each species is divided into fractions according to its percent of its food in that category. There are thus two ways of using such names as carnivore or herbivore—the one represented in Figure 3–2(c) puts a species in the category which dominates its nutrition; the other represented in Figure 3–2(e) puts parts of a species into different trophic levels. Either way it is a simplified model and what is meant can be made clear with the energy diagramming.

Finally, in Figure 3–2(f) compartmentalization can be further simplified into two sources, storage, and the process of production and respiration. For purposes such as estimating overall properties of the estuary, we often begin a study by evaluating the P and R compartmental diagram and then proceed toward adding detail, dividing categories as needed and as information becomes available. Figure 3–2(f) is the same as the diagrams in Figure 1–1(d) with exports and the detail on mineral nutrient flows grouped into respiration.

When we make groupings in network space, the geometrical positions may not be important as long as the connections of inflow and outflow are unchanged. If stretching the space of the system is not important, the phenomenon is said to be in topological space.

Traditionally, the environmental systems were studied in relation to ordinary geometrical space using maps, distances between organisms, densities, and other properties of ordinary geometry, sometimes named Euclidean geometry. Many of the newer studies that concentrate on circuit networks apply topological ideas when only the inputs and outputs matter. However, Euclidean space is also important; for example, plant chlorophyll is distributed relatively evenly on an area basis because the energy of the sun enters on a broad wavefront from above.

In defining the boundaries of a system, it is sometimes convenient to delimit a network and define its boundary-crossing circuits, thus making a topological boundary. More often, spatial boundaries are drawn as in Figure 3–1.

ENVIRONMENTAL SYSTEMS IN VERTICAL PATTERN

If we choose boundaries to include both production and respiration (P and R) functions, we often find that the functions are stratified in vertical layers as a response to the sunlight entering from above and the action of gravity in separating gas, liquid, and solid phases. The general patterns are drawn in Figure 3–3. The photosynthetic production and its chlorophyll-bearing structures become localized on top in the light, whereas the consumption parts of the cycle become concentrated below, the most remote states of the fuel flow often being lowest. The action of photosynthesis tends to deplete the raw materials for photosynthesis on top, releasing oxygen and accumulating new organic matter. The top zone receives the sharp diurnal and seasonal changes of insolation and serves as a roof, ameliorating the conditions of the system. The organic matter is carried downward by gravity or by the tube systems of tree trunks. Removal of the fuel to a lower zone leaves the upper production zone oxidized. The concentration of respiratory action below accumulates carbon dioxide, makes conditions acid, and releases again the mineralized critical nutrients needed by the plants for photosynthesis; this stratum is therefore called the regeneration zone. Oxygen tends to diminish because the respiratory metabolism dominates, and in some systems the oxygen reaches zero as other reduced chemical states develop, forming the reduced gases—hydrogen sulfide, ammonia, and hydrogen.

The various systems have some means for returning the regenerated nutrients to the photosynthetic production zone—through root transpiration in trees, through the vertical movements of the swimming or

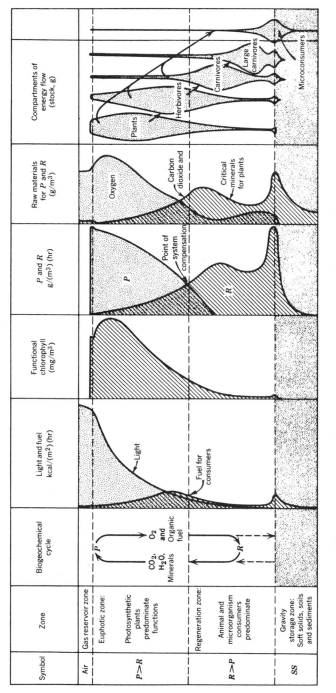

Figure 3–3 Principal vertical zones of many ecological systems on land and in water.

69

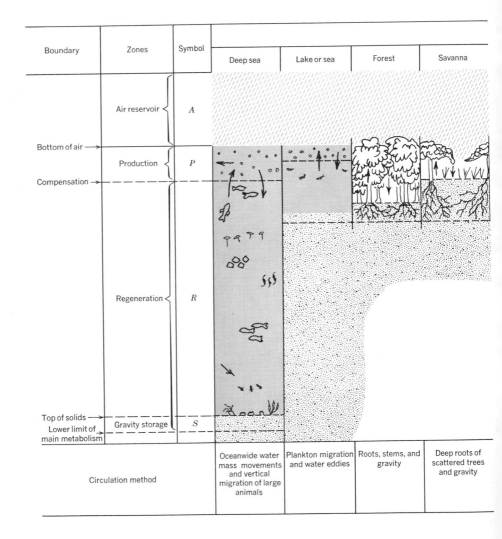

Boundary	Zones	Symbol	Deep sea	Lake or sea	Forest	Savanna
	Air reservoir	A				
Bottom of air →	Production	P				
Compensation →						
	Regeneration	R				
Top of solids →	Gravity storage	S				
Lower limit of main metabolism						
Circulation method			Oceanwide water mass movements and vertical migration of large animals	Plankton migration and water eddies	Roots, stems, and gravity	Deep roots of scattered trees and gravity

flying consumers who release wastes along the way, by release of gas bubbles that carry absorbed materials to the surface, and through other physical currents to which a particular system is adapted.

In Figure 3–4 ten systems with P and R zones stratified in the general manner of the graphs in Figure 3–3 are contrasted. The vertical scale ranges from a mile or more to a few centimeters. Some principal differences are due to varying vertical positions of the P and R zones in

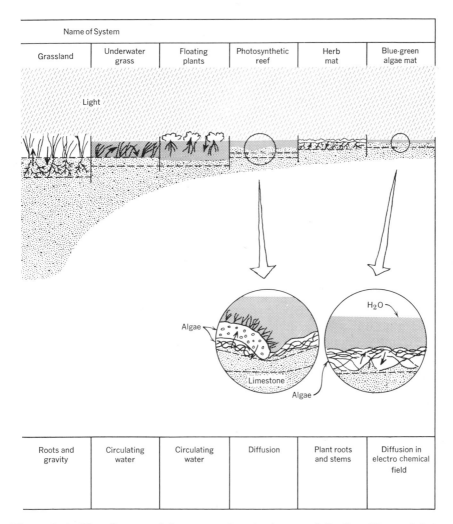

Figure 3–4 Identification of the principal vertical zones defined in Figure 3–3 in ten contrasting ecological systems.

relation to the gas-liquid-solid zones. In the thin aquatic systems the P and R zones are mainly within the porous solid phase. Another variable is the distance between the productive zone and the top of the solid phase on which falling organic fuels usable by the consumers accumulate. When the solid surface is distant, as in the deep sea, little organic matter reaches it and only after many years through indirect routes of world current and through successive handling by vertically migrating populations. When the mud surfaces are near the euphotic productive zone, they serve as centers of fuel accumulation and consumption. In general, in aquatic systems more of the circulatory work is done by fluid movements, whereas in land systems specialized plants carry on non-gaseous circulation.

HORIZONTAL PATTERNS

Production and respiration systems also have some common patterns when viewed horizontally. Because the incoming solar radiation is spread evenly on an area, the plant-producing tissue and its chlorophyll masses tend to be spread evenly. As we pass from forest to field, swamp, and aquatic medium, or to the agricultural systems controlled by man, there tends to be an even cover of green in spite of great differences in the kind of plants providing the support. The amounts of chlorophyll are different because conditions and the ratio of photosynthesis to chlorophyll vary, but many properties of the living cover vary much more.

Individual plant species are not evenly spread but are concentrated in zones according to their specialized adaptations to local conditions. More than one species is interspersed in the same kind of environment because networks need diverse pathways, a phenomenon discussed in later chapters. Random distributions are rare. Even though seeds may sometimes be spread by randomly operating dispersal factors, the factors that cause growth make the species distributions heterogeneous but the chlorophyll cover more uniform.

In contrast to the even carpet of green provided by the producer populations, the various consumers tend to be clustered. The large size of many animals tends to concentrate the respiratory function, with the single organisms collecting fuel over a wide area. Many consumers such as ants, antelope, and people have clustered populations because their complex functions are favored by grouping.

In Figure 3–5 the pattern of uniform production and clustered consumption is diagrammed. This pattern is observed in diatoms and tube feeders on glass slides that have been submerged in fertile streams for

Top view of *P–R* systems

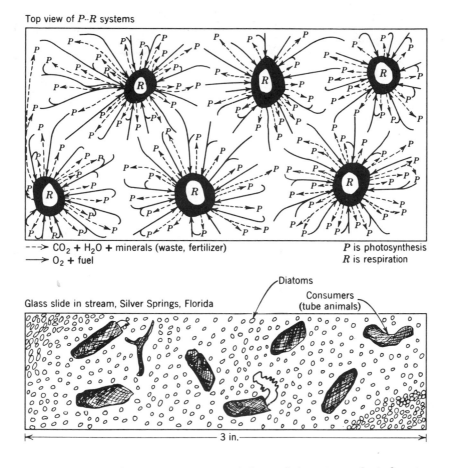

$--\!\!\rightarrow$ CO_2 + H_2O + minerals (waste, fertilizer)   *P* is photosynthesis
$\longrightarrow$ O_2 + fuel   *R* is respiration

Glass slide in stream, Silver Springs, Florida

Diatoms

Consumers (tube animals)

|← 3 in. →|

Figure 3–5 Frequently occurring patterns of metabolism in ecological systems viewed from above, showing productive photosynthesis dispersed evenly over the surface but the respiration clustered in centers and linked to production through convergence of circulating pathways.

six weeks. It is also observed from an airplane flying at a height of 10,000 feet over the wheat fields and farm villages of Kansas. The centralization of tree functions in a forest also follows the pattern, with limbs and roots spreading out from the trunk.

THE LIVING MASSES OF NATURE

Consider the living masses of nature's subsystems. Each system provides a fingerprint if a graph is drawn of the weights of its component groups—

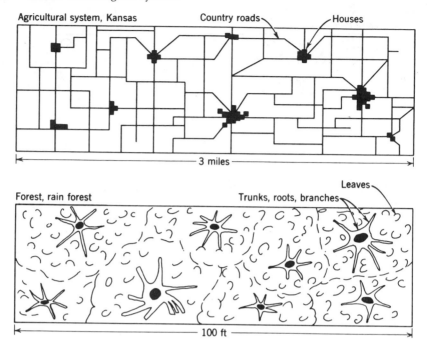

Figure 3–5 (*continued*)

plants, herbivores, carnivores, secondary carnivores, and fifth-order consumers—arranged according to the flow of the energy-bearing food.
Illustrated in Figures 3–6(*c*) and (*d*) are very different examples of
these graphs, sometimes called biomass pyramids.

Obtaining data on biomass representative of the large systems of nature
such as forests and seas is a very difficult task that has occupied ecological research for 50 years with thousands of methods and varying results.
When small spots are studied or small samples taken, the data are not
representative because of the large statistical variation that is characteristic of most ecosystems. Efforts to sample large sections of systems are
laborious and expensive. Figure 3–7 shows a helicopter-borne net, 50 feet
in diameter, which we used in shallow Texas bays to estimate the biomass of fish (Jones, Ogletree, Thompson, and Flenniken [12]. Consistent
with the ideal stated as a macroscopic view (Chapter 1), sampling was
done as though we were a giant far above the system oblivious of the
fine scale variation. Unfortunately, we had no giant.

At first the contrasting patterns of biomass seem to indicate vastly
different systems. The man strolling in a forest sees a great mass of plant
matter about him and a few relatively tiny small animals. Then later,

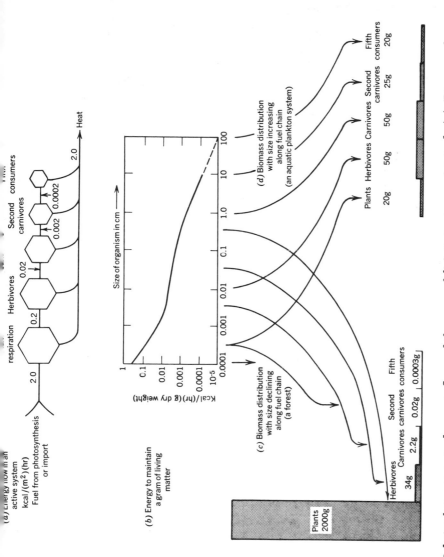

Figure 3-6 Relation between a steady energy flow and the mass of living structure maintained. (*a*) Diagram of a representative energy flow among compartments of the food chain. (*b*) Graph of the respiratory metabolism of organisms as a function of their individual size, based on data in Zeuthen [40]. (*c, d*) Two distributions of mass estimated by dividing the energy flow by the energy per gram maintenance requirements given in (*b*). Different sizes cause masses to vary, even though the energy flow is the same.

Figure 3–7 Helicopter-borne drop net ready for release of electromagnetic supports over Corpus Christi Bay, Texas. (From Jones, Ogletree, Thompson, and Flenniken [12].)

from his fishing boat in the sea, no plants are visible at all and an occasional fish seems the prominent and only part of the system. With such different proportions of foliage and creatures, we might conclude that the fundamental bases of these living patterns are dissimilar. Quantitative considerations show, however, that both follow from the same kind of energy flow because of a special property of living matter. For example, the biomass graphs in Figures 3–6(c) and (d) follow from the energy flow shown in Figure 3–6(a).

The amount of energy required per pound to maintain life becomes greater as the size of the organism decreases. An approximate relation between organism size and rate of fuel consumption is given in Figure 3–6(b). If we divide the calorie flow rate per compartment by the calorie flow rate necessary to support a gram of tissue, we obtain the grams of weight that can be maintained.

In Figure 3–6(c) organism sizes have been taken as diminishing along the food chain. A very steep pyramid like that for the forest results. Figure 3–6(d) organisms sizes have been taken as increasing along the food chain. The pyramid is very flat like that for the sea, and the total mass of the tiny plants is relatively small even though they process the most calories. These two illustrations are extreme; most systems have patterns that are intermediate.

In examining the distributions of living masses in the masses in the systems of the earth, we often find that the distinctively different graphs reflect similar energy bases. Biomass graphs are really pictorial inventory reports that can be used with circuit diagrams to describe fully a network and its functions. Although the examples given involve compartments of the intermediate view, biomass graphs by individual species provide a closer view.

Also convenient are graphical representations of the masses of non-living organic matter in peat, wood, litter, and sediment. These accumulations are part of the general system of storages that affect stability during fluctuating times, sustaining metabolism when inflows falter.

SEASONAL PATTERNS

The metabolic patterns for the systems of the earth have some general rhythmic properties that are caused by the seasonal pulses of dark and light in the program of the incoming solar energy. There is no place on earth that does not have at least a 25 percent seasonal pulse. In addition, the weather systems provide rain and stirring energy on a seasonal program. Some of the principal types of overall metabolic variations with time are given in Figure 3–8. Part (a) is a theoretical case studied in laboratories in which the conditions are held constant and the production (P) and consumption (R) remained fairly constant. Part (b) shows a seasonal oscillation attributable to variation in energy input, but the system by internal program mechanisms keeps consumption in phases with plant production. In (c) the two processes lag, one peak following the other, but both are rhythmic with the season. In (d) the P and R have large random fluctuations, although there is a fairly constant aver-

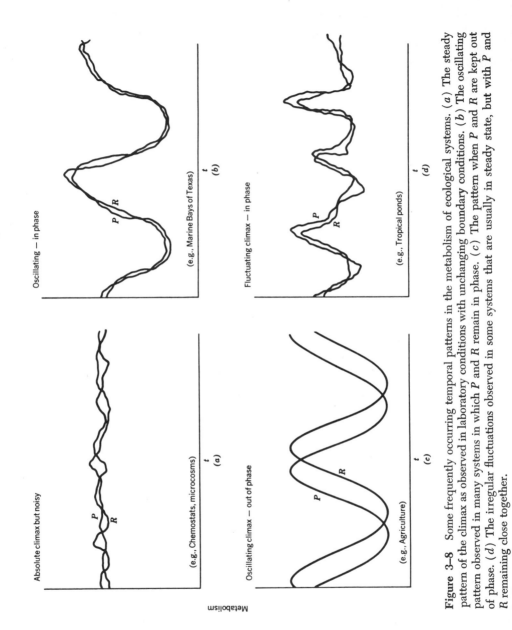

Figure 3–8 Some frequently occurring temporal patterns in the metabolism of ecological systems. (*a*) The steady pattern of the climax as observed in laboratory conditions with unchanging boundary conditions. (*b*) The oscillating pattern observed in many systems in which *P* and *R* remain in phase. (*c*) The pattern when *P* and *R* are kept out of phase. (*d*) The irregular fluctuations observed in some systems that are usually in steady state, but with *P* and *R* remaining close together.

age position when the system is considered over a long time, and when the two processes stay together day by day. Systems which have some constant or repeating patterns with time are called climaxes by ecologists. Four kinds are shown in Figure 3–8.

The various kinds of temporal patterns given in Figure 3–8, like the biosphere as a whole, have on the average a balance of P and R, although balance is achieved in different ways. Climaxes in which P and R are not equal are considered in later chapters.

COMPLEXITY OF NETWORKS

In a close view of a system, we may measure its complexity by counting the number of species and occupations. The number of routes of energy and material flow is proportional to the kinds of functional units (the species and occupations). Much of the science of environmental networks is concerned with the patterns in these numerical data. In Table 3–1 are some data on network complexity obtained by counting the numbers of functions represented in a sample of 1000 individual functional units (organisms and people).

Table 3–1   Species and Occupational Diversity in Some Systems

System	Roles Found when 1000 Individuals Are Counted[a]
Systems with few species in stressed environments	
Mississippi Delta, low salinity zone	1
Brines (Nixon, [16])	6
Hot Springs	1
Polluted harbor, Corpus Christi, Texas	2
Humans on a military frontier	
(Infantry Table of Organization)	10
Complex systems in stable environments	
Stable stream, Silver Springs, Florida	
(Yount, [39])	35
Rain forest, El Verde, P. R.	75
Ocean bottom[b]	75
Tropical sea[b]	90
Human city (see Fig. 4–11)	111

[a] When fewer than 1000 are counted, statistical agreement between duplicate counts is not good; when larger numbers are counted, work is unduly tedious and discontinuities between different systems tend to be crossed. All units in the same size realm are counted.

[b] Sanders [32].

When ecological systems must adapt their functions and organs to survive in conditions that are physically severe for life such as extreme heat, cold, or saltiness, energy is required for the processes of adaptation and less energy remains to support complex network specializations. There are fewer species and occupations. The first examples in Table 3–1 are severe environments with few species or occupations.

In contrast are the last examples of Table 3–1 which have more specialized functional circuits. The large and concentrated flows of energy into an industrialized city support extreme specialization of occupational functions not possible for the same number of people in an Eskimo village. Similarly, the rain forest has more circuits than does the brine. The complexity and diversity of the macroscopic world are thus related to the flows of energy and the budgeting of these flows among various functions.

MAJOR SUBSYSTEMS OF THE PLANET

In the grand view of the biosphere the environment has many spatial subdivisions: seas, tundras, deserts, and great rivers. Several generations of observers have felt that division boundary lines could be drawn to group together terrestrial systems with similar structure and function. Thus, the areas of coniferous forest of the colder regions have been grouped together, even though the conditions and actual species of plants and animals constituting the system on one continent are somewhat different from those on another continent. Ultimately, zones occupied by the major subsystems may correspond to climate zones which determine the amount of available energy and the energetic costs of adaptation to desert, temperate, tropical, and arctic conditions. In Table 3–2 a few classical "plant formations" are included in the classification of subsystems of the earth using major proportions of its energy budget.

A classification of this planet's systems is, however, based on other factors than zones. Systems utilizing solar energy cover a large area, but the industrialized city draws a large energy flow without covering a large area. Functionally the city is as much a major subsystem as the forest.

The circulation pattern of the sea has gyres, loops, and other means that allow characteristic biological communities to remain in about the same place, even though the water is in continuous motion and ultimately intermixes. For example, the plankton systems in the center of the subtropical oceans are remarkably stable in structural pattern. Since a great part of the energy budget of the earth goes through the aquatic medium, some of the larger aquatic systems must be included in the listing of major planet subsystems sometimes called biomes.

Table 3–2 Some Major Classes of Ecological Systems of the Earth Arranged from Pole to Equator

Name of System Class	Example	Characteristic Energy Source[a] Stress, or Program
Land		
Tundra	Pt. Barrow, Alaska	Permafrost
Northern coniferous forest	British Columbia	Heavy snow, cool summer
Deciduous hardwoods	Eastern United States	Cold winter, hot summer
Prairie	Kansas	Limited rain, fire
Temperate desert	Utah	Limited rain, cold winter
Subtropical desert	Northern Mexico	Erratic rain
Savannah	Central Africa	Seasonal rain, deep roots
Deciduous tropical forest	Panama	Sharp dry season
Tropical rain forest	New Guinea	Even rain
Marine		
Arctic iced seas	Bering Sea	Mainly ice-covered, sharp season
Temperate-plankton-herring sea	North Sea	Temperate climate
Low salinity system	Baltic Sea	Low salinity
Estuarine plankton	Chesapeake Bay	Salinity fluctuation
Consumer reef	Oyster reefs, Texas	Current; salinity variation
Kelp bed	California	Cold, clear shores with swell
Marine meadows	South Florida	Shallow banks
Mangroves	Puerto Rico	Tropical shores
Blue water tropical seas	Caribbean, Sargasso Sea	Low nutrients
Upwelling zone	Peru	Enriched nutrients
Coral reef	Australian Great Barrier Reef	High-energy tropical shores

[a] The special factors listed affect energy by controlling limiting factors or stresses.

In naming subsystems of the earth it is customary to choose some physical feature or population that is most prominent and may be said to dominate the system. The relative dominance can be quantitatively indicated by the energy flow involved.

We should hasten to indicate that many of the most fascinating subsystems are small, occur infrequently, and at present involve little of the overall budget of the earth. Although not listed in Table 3–2, these are no less important than the big divisions because the little ones may represent systems that dominated in past ages, systems that will cover the earth in the future, systems that help us visualize possibilities on other planets, systems that can be engineered for the benefit of man, and systems that have great value for future scientific progress.

MAGNITUDES OF ENERGY FLOW IN NATURAL SYSTEMS

Enough of the natural systems of the world have been studied energetically to establish the probable magnitudes and ranges of energy flow mostly as photosynthesis. Some of the highest and lowest values are summarized in Table 3–3.[1] The energy flows vary daily and seasonally since the input energies vary in a similar manner. Generally speaking, the seasons of bright light are shorter in the temperate latitudes than in the tropical areas so that the energy input and the power flows through the ecological systems there (1500–4000 kcal/$(m^2)(day)$) are in a ratio of about two to one (tropical energies, 4000–5000 kcal/(m^2) (day)). In the polar areas the energy input and power flows are smaller (2000 kcal/$(m^2)(day)$), although very intense for the short summer period.

The data selected for Table 3–3 cover the world's range of primary production. Listed first are low values characteristic of both land and water in much of the subtropics where there are shortages of critical raw materials causing recycling mechanisms to develop. Listed next are systems that are cultivated for maximum photosynthetic production by

---

[1] Production can be measured as oxygen released, carbon taken up, weight increased, and so on. In the chain of biochemical steps, oxygen is the first readily measurable by-product which is released by the plant cells after light energy is incorporated as chemical potential energy. Other substances that are taken in (such as carbon dioxide, nitrogen, and phosphorus) are involved further downstream in the reaction chain after energies are beginning to be dispersed in work and after flows are distributed in branching routes and time-delaying storages. Thus, oxygen gives higher measurement estimates of photosynthetic production. Most systems have concurrent respiration in the day in the same cells, using much oxygen that would have emerged. No good way has been found to measure this. The respiration at night is not a good measure because it is often only a small fraction of that during the day.

Table 3-3   Magnitudes of Primary Production[a]

System	Rate of Production of Organic Matter (kcal/(m²)(day))[a]	Efficiency (%)[b]
Systems with little measurable production (most photosynthesis either masked by internal cycling, lags, and storages, or nonexistent)		
Subtropical blue water[c]	2.9	0.09
Deserts[d]	0.4	0.05
Arctic tundra[d]	1.8	0.08
Fertilized systems with high production that is stored in seasonal growth, organic storage, or outflowing yield		
Algal culture in pilot-plant scale[e]	72	3.0
Sugar cane[f]	74	1.8
Water hyacinths[g]	20–40	1.5
Tropical forest plantation, Cadam[h]	28	0.7
Systems with measurably high production which is mainly used the following night (minerals being recycled, P and R similar)		
Sewage pond on 7-day turnover[i]	144	2.8
Coral reefs[j]	39–151	2.4
Tropical marine meadows[k]	20–144	2.0
Tropical rain forest[h]	131	3.5
Waste-receiving marine bay, Galveston, Texas[m]	80–232	2.5
Clear spring stream with vegetation-covered bottom, Silver Springs, Florida[g]	70	2.7

[a] Day net production plus night respiration. This procedure omits much immediately used photosyntheses. For more detail data see general reviews by Pearsall [28], Ryther [30], and Westlake [38]. Data that were originally given in grams of oxygen were converted to kilocalories using 4 kcal/g. Data originally given in grams of carbon were converted using 8 kcal/g.

[b] Percent of total sunlight received. About half of sunlight is directly involved in photosynthesis but the heat radiation received may also contribute to such auxiliary processes as evapotranspiration, photorespiration, and rates of biochemical reactions.

[c] Ryther [30].

[d] Rosenzweig [31].

[e] Tamiya [34].

[f] Odum and Odum [25].

[g] Odum [18].

[h] Odum and Pigeon [26].

inflows of fertilizing nutrients. These values are 100 times higher than those of the limited areas. These systems provide storages and yields for man. Finally, in the third listing are some natural systems well organized for effective recycling of minerals. Such systems replenish their needs by regenerative respiration. These natural systems have photosynthetic rates as high as the fertilized-yield systems. Man's efforts to augment food production by plants rarely exceed the production of well-organized natural systems, although man channels it into human use rather than to the animals and microorganisms.

Ecosystems in regimes with sharp seasons tend to store production during their growing periods to last during their fallow periods. These systems demonstrate their productivity by their storages. Systems in steadier climates tend to use their production as it is made and the basic productivity is not obvious since there are no great accumulations. Daily gains can usually be measured if oxygen is released or carbon dioxide taken in. As yet there is no easy way to estimate the basic photosynthetic process including the organic matter that is used as fast as it is made. Because of this difficulty, there are still many doubts about the distribution of the basic photosynthetic process over the earth. Respiratory consumption in many environments is accelerated during the daytime with the aid of light and the impetus of fresh photosynthesis which fills internal reaction pathways. The tendency to utilize what is produced in tight internal cycles is greatest in the areas where some material requirement is in short supply as is water in the desert where plants recycle or chemical fertilizer elements in the blue tropical seas where phytoplankton recycle. These areas may have high productivity without very many depositions or gaseous exchanges outside of the green plants. The maps of the world's productivity so far are based on net storages and gas exchanges, but they serve a purpose in showing the areas where there are net foods available to other organisms. They do not as yet portray the distribution of total power flow in the biological systems.

CONCENTRATING THE ENERGY OF THE SUN

The input of dilute sunlight falls on leaves or on distributions of planktonic plant cells in water, and photosynthesis occurs. The stored energy of organic matter produced over a broad surface at a slow rate is then

[i] Ludwig, Oswald, Gotaas, and Lynch [14].
[j] Helfrich and Townsley [11].
[k] Odum [23].
[m] Odum, Cuzon, Beyers, and Allbaugh [22].

collected and concentrated by the consumer systems of animals and of tree twigs and limbs. The cost of the concentrating work is paid for from some of the collected food. This system is possible only when the power drain costs of the collection work are less than the power input achieved by the collection.

Figure 3–2(b) summarizes the energy concentrations involved in the flow of power through the stages of an ecological system as the total power in available form decreases. The protein content and other aspects of quality of the organic matter increase with the concentrating process. By combination the trend in nutritional quality is toward chemical diver-

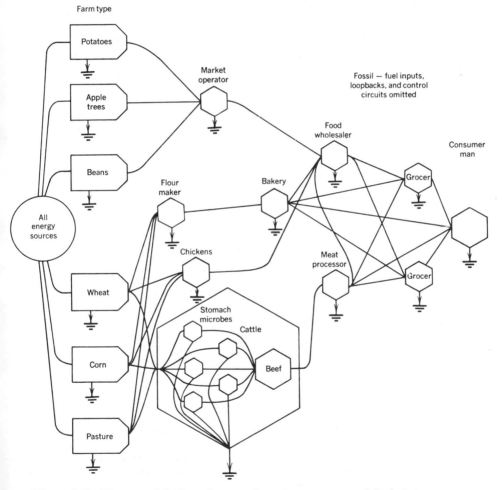

Figure 3–9 Diagram of the loss of power through convergence of food chains necessary to develop quality nutrition.

sification and toward the more exact composition of the body structures of complex higher animals, containing proteins, vitamins, and so forth. The diagram emphasizes the high cost of such synthesis in power dispersed through useful work. Figure 3–9 shows some of the converging pathways by which high-quality meals reach our table in an industrialized society. Notice some of the same tendencies to lose energies in the collecting and processing as in the primitive system. For the industrialized system the units are aided by fossil fuels. Without this supplement 99 percent of the original food is used in concentration and increasing quality.

The energetic cost of the special, concentrated, and ordered chemical factories in tissues of higher organisms is very large if we include the costs of supporting the multiple species networks and many specializations necessary to keep the ecosystems running. These relations are often described as a pyramid of power flow, with very little reaching higher stages and thus with little possibility of supporting large populations at

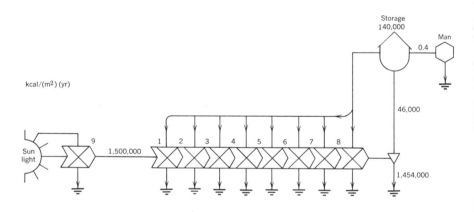

Figure 3–10 Work processes in a surviving system; a summarizing energy diagram for forest man and the tropical rain forest system (Odum [24]). See Figure 3–2 for the type of network for which all the details are rarely known. The multiple-species networks of food chains and mineral regeneration routes accomplish the following work flows: (1) regenerate plant materials; (2) provide epidemic protection with special biochemical substances in each plant species; (3) limit any one microorganism or insect species by generalized carnivores and omnivores; (4) provide continous chlorophyll receptors for maximum use of light; (5) provide stable programs of fruiting and other reproduction; (6) gather and converge plant-specialized nutrition into higher-organism meat; (7) provide shade to control ground invasion; and (8) maintain soil structure with organic matter, burrowing animals, roots, and microbes. (9) Auxiliary energies from the action of winds, rain, and other flows of the biosphere contribute to system function.

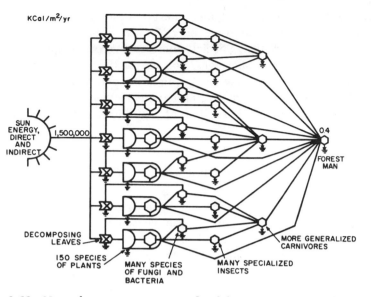

Figure 3-11   Network matrix supporting and stabilizing a tropical rain forest system. Man is a minor component, but he has integrating and control functions because of the convergence of pathways [24].

that level. Compare the new human economy sector in Figure 3-9 with older patterns in Figures 3-10 and 3-11.

In ecological systems a number of necessary functions are accomplished within the organisms, within the population group activities, or within the interactions among populations. We may summarize some of these activities with the work diagram in Figure 3-10. As yet, the proportion of the system budget normally spent on each is not known. If any essential function is omitted in one population, some other population may have greater staying power and may replace the faulty one. Because a system and all its parts are continually putting out offspring through reproduction, including new experimental variations, there is always a series of alternative possibilities for the continuing network. Using Darwinian language, we say that nature selects from the choices to determine what predominates thereafter.

Essential functions of the system are reproduction of the damaged parts and rearrangement of disarranged aspects. Some energy must be diverted to new evolution, as described in a later chapter, and some energy is used by mechanisms of self-switching. Some energy must go into storages which may be necessary for structure or for smoothing out fluctuating power supplies. Some energy must go into the work of gathering input flows, and some energy must insulate and control the relation of other circuits. Special energies overcome special limiting factors, such as material shortages, poisons, or the seasonal fluctuations. Finally, special energies may be required to pump out excess wastes. Different species have different combinations in their programs of spending their energy budget. For different conditions different species are best adapted.

ORGANIZATION IN THE COMPLEX RAIN FOREST

The energy diagram in Figure 3–11 summarizes the network of many species that is found in the stable, complex communities in which man has possibly evolved as a general organizing component in small numbers. Such systems are found where the climate does not have a sharp pulse in drought, temperature, or other factors that force a system to start over each year. Man is found in low densities of about one per square mile in many systems.[2]

Some of the system's principal needs for survival and stability are provided by the arrangement of flows shown in the network in Figure 3–11. The following theoretical discussion of the way this network may operate is consistent with observations of the most important species in a rain forest at El Verde, Puerto Rico. Photosynthesis (gross) there is about 32 g dry matter/(m^2)(day). First, the plant photosynthesis is divided among many different species, each of which has a similar photosynthetic function but is made distinctive by special chemical insulating agents. The cycle for discarding old leaves and recycling minerals is provided by fungi and bacteria with considerable specificity to deal with the special insulating agents. Various arrangements of roots and microorganisms at the ground level further organize closed-loop mineral cycles which are mutually stimulating.

Both the producers and the microorganisms are regulated by specialized consumers. Leaf-eating insects are chemically specialized to deal with the plants, but they are adjusted in their actions to take normally no more than about 7 percent of the leaf matter, as long as each plant

---

[2] Hagen [10] gives a population in 1940 of 1.4 persons per square mile for the Amazon, including the towns. Turnbull [35] gives a population of 40,000 pygmies for a rain forest area of 50,000 square miles, or 0.8 person per square mile. In Australia, Birdsell [2] found aboriginal populations in densities up to 550-person villages per 600 square miles in high rainfall areas and down to 550 persons per 40,000 square miles in dry central areas.

species is evenly dispersed. Similarly, a great many species of small fungus-eating flies and other small consumers may keep the microorganisms from becoming too numerous. Both the leaf-grazers and the microbe-eaters are specialized to deal with the chemical insulating compounds of plants. As long as the various plants are dispersed in relatively small numbers for each species, the plant- and microbe-eaters cannot find enough continuous specialized food to become epidemic. If any species of plant starts to accelerate its relative growth and become dominant, it will be counterattacked by accelerating actions of the consumers whose small size gives them a rapid response time. Thus diversity and a check-and-balance system prevent epidemics and maintain stability, a theory for which much evidence has been gathered by Vouté [37].

Higher-level consumers, such as the lizards and frogs, and secondary-level microorganisms and eaters of detritus, such as the earthworms, need less chemical specialization because they are not eating the organisms with special chemical insulators. Instead, they are eating the small insects which, although specialized in function, are less so in their own chemical machinery. Thus a convergence of food lines is possible, with the higher consumers serving to shift from one species to another if either becomes too numerous. The convergence also provides a variety of chemical molecules, supplying more complex nutrition which, in turn, allows for the more complex functions and tight packaging required for the rapid movements, elaborate behavioral adaptations, and diversity of the higher organisms. The specialization in higher organisms permits complex patterns of system control such as control of pollination and seed fall.

Ultimately this convergence leads to larger birds, mammals, and man, providing nutritive diversity and a stable but small flow, for no one chain can be augmented without being counterbalanced by the control system already mentioned. Thus the energies of photosynthesis go into hundreds of different species circuits and special work processes by which the overall needs of the system (Fig. 3–10)—mineral cycling, epidemic stabilization, structural organization, effective light utilization, maintenance of soil structure—are achieved.

The rain forest system may be drawn with different compartmentalization as shown in Figure 3–12 which has the feedback of work services of downstream consumers contributing to the upstream by regulatory work. Small amounts of leaves (7%) are taken by herbivores in normal times.

POWER TRANSFER EFFICIENCIES

The various forests, reefs, and natural algae-based ecological systems have been evolving for many millions of years and perhaps it is not

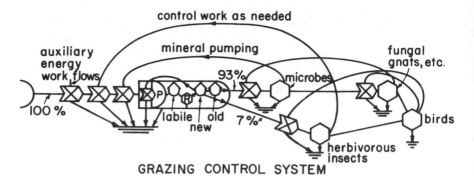

Figure 3–12 Compartmentalization of a rain forest system to show the roles of herbivores acting on leaves, the microbes on dead leaves, animal managers of microbes, and the overall switch-feeding control action of birds (Odum and Pigeon [26]).

surprising that the criteria of importance to survival have been developed to their maximum by the evolutionary process. Because every real process requires power, the maximum and most economical collection, transmission, and utilization of power must be one of the principal selective criteria as Lotka stated in 1925 [13]. Many of the contrasting systems of nature on similar input energy budgets find their maximum development and maximum power flows approaching similar values, suggesting that some kind of energetic limitation has already been achieved. These limits are of extreme importance to any consideration of the world's capacity for sustaining man now and in the future. Sometimes it is useful to refer to the concept of efficiency in describing the amount of useful power transmitted after the first, second, third, or fourth energy transfer stage.

Referring to Table 3–3 we find the efficiency of gross plant production to be about 2 percent and that of transfer to the animal consumers (herbivores) to range from 0.5 to 49 percent (Table 3–4). The high efficiencies of growth by the steer under fattening diets occur when the animals are not running about doing their own support. Some of the food substances may be digested, assimilated, and deposited without much metabolic processing.

The marine zooplankton (Table 3–4) when kept in laboratory situations with dense food concentrations have efficiencies almost as high as that of the steer. At the other extreme is the elephant on its wild African range with only a small fraction of its assimilation going into weight gains. In many studies of wild animals in nature, figures for efficiency are intermediate, in the vicinity of 10 percent of intake as weight gains.

Table 3–4 Examples of Efficiencies of Food Chain Transfer in Plant-Eating Animals[a]

System	Ingestion into Assimilation (%)	Assimilation into Growth (%)[b]	Overall Ingestion into Growth (%)[b]
Symbols in Figure 3–13	P_2/P_1	P_3/P_2	P_3/P_1
Steers under fattening diet[e]	66	74	49
Marine zooplankton in rich laboratory conditions[c]	59	57	34
Elephant on natural range[d]	32	1.5	0.48

[a] Animals include populations of stomach microorganisms that eat and process food yielding nutritive substances to host.
[b] If population is exporting (P_3) at the same rate that there is new growth, the population is at steady state.
[c] Means of medians of 10 groups of data from literature and 12 other experimental studies (Conover [6]).
[d] Petrides and Swank [29].
[e] Brody [3, p. 82].

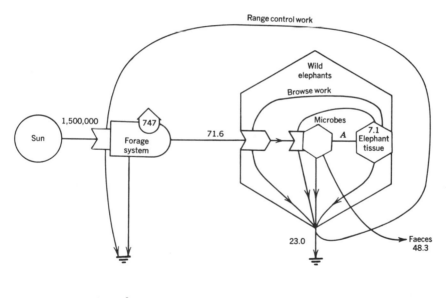

kcal/(m²)(yr)

A = Assimilation

(a)

Figure 3–13 Some of the main energy flows within a self-maintaining population represented in energy diagrams with the hexagonal symbol. (a) Elephant populations in Africa (data from Petrides and Swank [29]); (b) component flows of a population with environmental control of seasonal program of reproduction.

91

Figures for elephants, cows, and most herbivores really apply to two stages of food processing—first to the microorganisms in the elephant's complex stomach and then to the elephant (Fig. 3–13(*a*)).

Any ratio of power flows is an efficiency and there are as many kinds of power ratios as there are pathways. Many names have been used to describe various kinds of efficiencies, and definitions are not always clear. Energy diagrams help. Figure 3–13(*b*) shows some of the energy pathways within one species population as food flows in to accomplish various kinds of useful work and leaves the population as wastes, heat, or loss of members. The figure indicates the details implied in the use of a hexagon-shaped symbol for self-reproducing entities such as cells, organisms, populations, consumers systems, and cities. The efficiencies for

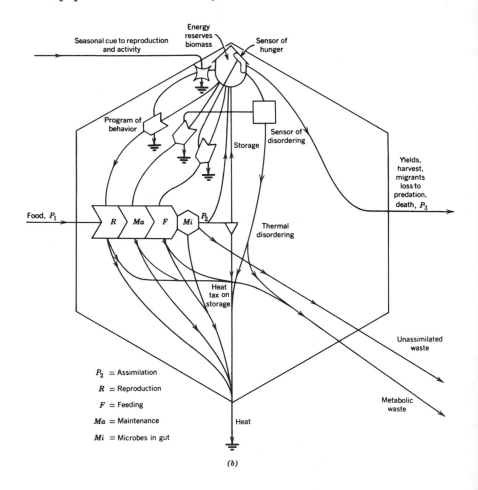

P_2 = Assimilation

R = Reproduction

F = Feeding

Ma = Maintenance

Mi = Microbes in gut

(*b*)

conversion of input food into new organic matter for other consumers in Table 3–4 were calculated as the ratio of P_2 and P_3 to P_1 in Figure 3–13(b). If the yield is not taken, the stocks and system structure develop. When yield is taken from a virgin stock, the maintained levels decrease. There is an optimum removal rate that optimizes the stock to its energy

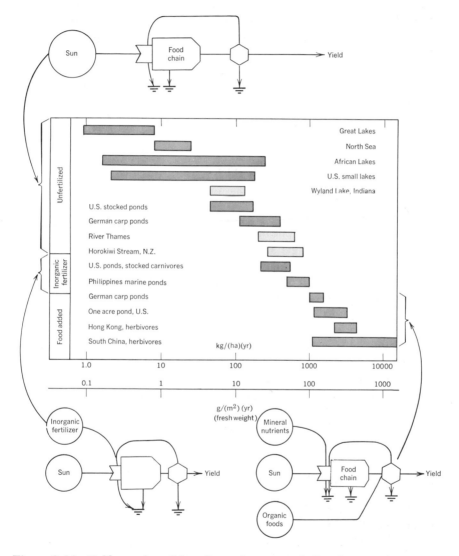

Figure 3–14   Yield rates from fish, redrawn from Mann [15] with appropriate energy diagrams.

input with the maximum steady drain of the yield. This is called the optimum catch.

Sometimes it is easy to compare meat production directly to the eco-system energies. Long studied and tested are the fish pond technologies which help us establish the levels of food yields that are possible for various energy inputs. In Figure 3–14 are representative data on yields with three energy situations shown by the energy network diagrams. The first group with only sun yields up to 4 g dry weight/(m^2)(year) depending on natural fertilizer conditions. With the energy of inorganic fertilizer added (see Chapter 2 for an explanation of limiting factors as energy contributions), yields are higher, and in the third group with organic foods added that feed the fish more directly yields go to 200 g dry weight/(m^2)(year). These data are useful in evaluating potentials for world aquaculture.

POWER AND CONTROL CIRCUITS

In very highly organized natural systems the flows of power are much divided among the species circuits, but they can be roughly separated into power circuits and control circuits. Thus, if oak trees process 50 per-cent of the power budget of a forest system, they constitute a power circuit. The squirrels of that forest may be processing much less than 1 percent of the forest budget. Their procedures for gathering and plant-ing acorns may, however, serve as a control on the patterns of the oaks. Thus, we must distinguish between the power flow in a circuit and the power being controlled by a circuit. The metabolism of managers in a large industry is small, but through electronic devices and personnel directives they control much of the vast power flow. Power circuits must be large and sluggish, whereas control circuits with small energies are easily insulated and can perform delicate operations and provide a direc-tive influence on the power circuits. These principles are the same in electrical power distribution, in the forest, or in the complex industrial systems of man.

NATURAL HISTORY AND POWER CONTROL

In the ancient lore of natural history we learn about the animals and plants and their life histories, when they breed, how they live and inter-act, and what all the little aspects of their behavior are. The emerging new sciences of behavior document the facts in a systematic way and deal with the various kinds of machinery and operation that cause this behavior. Thus a male bird with feather movement 1 sets off response 2 in the female which sets off circuit 3 in the male, and so on, continuing as keys in locks until the function is performed. The process is like that of a treasure hunt for which we must first fit together the pieces of a map; the treasure is some complex function such as reproduction and nurture. The sequence of steps forms the same kind of flow chart that a computer programmer draws in explaining the successive steps of a program.

What may be overlooked is that the higher animals and to some extent the plants that have complex behaviors and life cycles, all with vast power requirements, were not developed for man's enjoyment, by accident, or even through some quirk of evolutionary procedure. The power demands of these processes were reinforced by the system in which they developed and survive because these complex behaviors serve as control programs for the system. Thus the complex migrations of crab, shrimp, mullet, and other fishes from the Atlantic coastal bays and estuaries out to sea in the winter and back in during the spring provide a means for rapidly reinjecting a consumer system as the power input of the sun increases with the warmer weather. Without the migration there would be a lag with much input power lost. The seasonal pulse of production, total respiration, and migrating stocks in the shallow Laguna Madre of Texas is shown in Figure 3–15 [23].

BEHAVIOR AS SYSTEM PROGRAMMING

The eccentric behavior of the birds, bats, and blooms is really the outer manifestation of preprogrammed computer units that control the timing for the whole ecological system. The present science of behavior often asks only half of the why questions, namely those concerning the mechanism within the organisms for achieving the observed animal activities. Behavioral science will immediately become of age and make enormous strides when it also asks and measures the role of the species in doing jobs for the system. The huge variety of known species[3] is like a bin of electronic parts, tubes, transistors, resistors, and relays. Just as the properties of electronic parts are listed, tabulations of the performances of the species are needed, including information on such things as storage capacity, power demand, cycle frequencies, number of input plugs, etc.

[3] The possible use of rare species as special parts justifies considerable expense in preserving them from extinction since their original development costs were large and spread over evolutionary time.

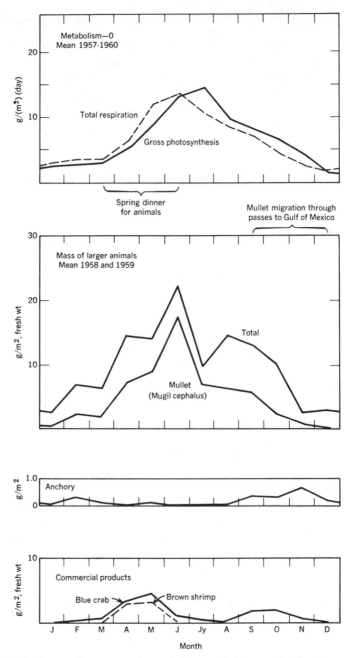

Figure 3–15 Seasonal patterns of metabolism and biomass in the Laguna Madre, a Texas estuary system. Population data were supplied by Thomas Hellier, Arlington College, Texas.

The design of existing ecological systems will then be better understood and planning for future designs can begin. In a similar way, the various occupations and operational groups of man constitute a bin of parts for the design of societal systems, and the same methodology can be applied in their study and manipulation.

COMPLEX AND BEAUTIFUL SYSTEMS

Nature reaches its most appealing manifestations of beauty, intricacy, and mystery in the very complex systems: the tropical coral reef, the tropical rain forest, the benthos-dominated marine systems on the west coasts of continents of temperate zones, the bottom of the deep sea, and some ancient lakes of Africa. These places are environmentally stable, and the species networks that develop there have great specialization and division of labor in the same way that the city of New York has enormous diversity of occupations. Complex networks must maintain controls for stability, but there is energy for control because the environments are initially stable, requiring no great energy drains for physiological adaptation.

The great diversity of species really constitutes an invisible network of pathways for food and materials in the populations. This network is comparable to complex electrical circuits which not only have their wires insulated against the short-circuiting of flows but also have the wires color-coded so that the organizer (in this case the electronics technician) can keep track of the circuits. In the complex species systems bright colors also develop, at considerable biochemical energy drain for the pigment syntheses. These colors and the behavior characteristics permit recognition when the mental powers of the populations are not large. Many of the bright colors are not present for beauty or as an accident. Color-coding of electronic circuits and color-coding of species populations may be essentially similar functions. The cities also use color to help organize communication.

In rain forests where there are great trees of much diversity, the circuits of food making and mineralization are facilitated through microorganisms. By making special biochemical compounds to differentiate each species from the others, each circuit is insulated to encourage specificity in microorganismal consumption, and so forth (Fig. 3–11). Recent research has proved the great specificity of these compounds. Those for each species of plant are quite different, and the enzyme systems of plant consumers to make the loopback of mineralization are correspondingly dissimilar.

The many channels required to make the special chemical and physical structures for maintaining ordered networks take much work and therefore much metabolic energy. Thus we find a very high respiratory metabolism in the complex systems, although the energies do not go into net storages and gains. The power budgets are among the highest since specialization allows the whole network to excel over alternatives, but no one product or species receives much energy and what each unit obtains goes into the work for the system rather than into net piling up of offspring, wood, coal, or other products. Complex systems may be very efficient in utilizing all the organic matter they have, leaving little to the fossil record. If this theory is correct, the great fossil coal beds were not derived from complex systems.

SIMPLER SYSTEMS IN SEVERITY

In many other environments of the earth the conditions for life vary sharply with the day, with the season, and from year to year. The estuaries of Texas are salty in years of land drought and then are suddenly flooded with freshwater regimes when the rains come. The Mississippi River fluctuates in its pathways to the sea, leaving bays saline at times and fresh in other years. Sharp seasonal changes occur in the polar tundra, in the deserts, and in the waters that receive the fluctuating wastes from man's sewers. Great energies go into survival of life at a minimum and into the constant replacement of decimated stocks. Little energy finds its way to diversification and speciation. Species must be generalists and acting as a group do not process energies very efficiently. More kinds of functions and programs are required in the body's physiology. Physiological diversity replaces species diversity. Large net yields of matter are frequently the output rather than work spent on specialization. The patterns of color, visual cues, and beauty are lacking, but nature is exciting in another way, for the few dominant stocks of creatures are characterized by great blooms, pulses, crops, and migrations. Man can secure a quantity of food from this kind of system, but only irregularly. Figure 3–15 provides a picture of the annual pulses of the estuarine network, similar to that in Figure 3–2.

OSCILLATORS

The simple ecosystems are often observed to oscillate. The famous much-studied examples are the cycles of abundance and scarcity of arctic animals, jackrabbits, lynx, and snowy owls. The systems models help us understand the mechanisms of these patterns. The Lotka-Volterra theory long ago showed the way a lag in the eating and being eaten of a prey-predator sequence could give undulations in the stocks. Many demonstrations of this principle have been made in laboratory populations in simplified situations in which there were no additional controls. Examples of such an oscillation are given by Drake, Jost, Frederickson, and Tsuchiya [9] and Utida [36] (see Fig. 3–16). The model for such oscillations is widely known in other fields.[4]

The discussions of these oscillations often suggest that the system being simple cannot stabilize itself with control mechanisms. However, relatively minor changes in timing factors can shift an oscillating system to a steady one.

If the selective factors in survival are keeping the ecosystem oscillator operating in this oscillator mode, there may be adaptive value. In electronics, oscillators are used to gain stability, the repeating pattern of oscillation being more easily regulated than a steady flow. Perhaps when systems are simple (few species) the oscillator mode is the easiest way to gain stability, especially when the environmental pattern is itself receiving the annual undulating patterns of the climate. When oscillations are a device for survival and when they are a detriment is very much an open issue in ecology.

OIL AND LOW DIVERSITY IN FOSSILS

When there are only a few poorly organized and little specialized species, organic matter is not efficiently utilized and relatively large quantities go into storage in the sediment and fossil deposits. Thus the oil and coal

---

[4] In systems engineering a chain of two storing units (integrators) with a loopback is called a second-order system because the differential equation for the whole system has a second derivative. Such systems may oscillate regularly, and when the input is variable, undesirably wide fluctuations may result. The oscillator may get into phase with the external variation of input and go into resonance with variations in storages and flow that are larger than those in the input. One example is a chain of two operational amplifiers in a standard analog computer. Patten [27] pointed out that such arrangements are used as sine generators. Sine generators are means for sending out an undulating, up-and-down voltage that follows the position of a ground shadow projected by a nail rotating on a wheel on a stationary axis. The sine generator simulates approximately the energy from the sun as it rises and falls during the day or by season, because of the trigonometry of the angle at which it strikes the earth's surface. Patten suggested that the sine generator is a possible model for the oscillations of two animal populations in a chain, such as a prey and predator relationship. Any population serves as an integrator summing food incomes and energy drains in its own systems network.

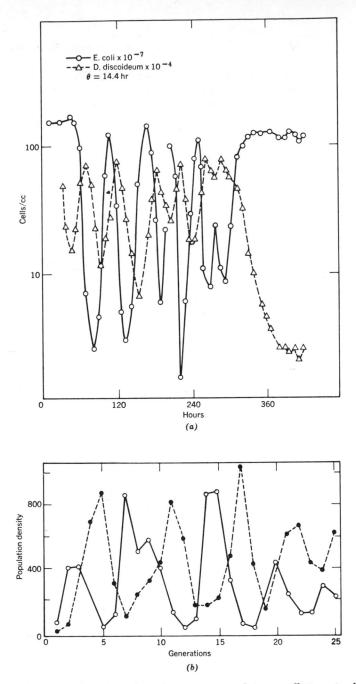

Figure 3–16 Example of Lotka-Volterra prey-predator oscillation in laboratory culture. (*a*) Protozoans eating bacteria (Drake, Jost, Frederickson, and Tsuchiya [9]); (*b*) grain beetles and wasps (Utida [36] from Browning [4], p. 78).

deposits may be related to conditions of severity or fluctuation. Many of the sedimentary areas of the world carry no oil because they had stable conditions and living networks that were efficient. Recognition of this simple principle might have saved vast amounts of money spent on oil exploration, for we can count the species diversity from the little animal skeletons in the drill cuttings of exploratory mining. Possibilities for finding oil are predicted when the species diversity is found to be less than 5 species per 1000 animal skeletons counted because such systems have been observed with high levels of organic matter and residues in their sedimentary depositions.

These and many other cues help us interpret the ecology of the past. As time goes on we see how wide and general was the storage of organic matter in the rocks of the past. Sometimes in half-seriousness we say that man may have been evolved by the system as a mechanism to get the fossil fuels and other minerals back into circulation. We hope he is pre-adapted for other roles after that.

SUMMARY

Following a billion years of evolution, our biosphere is populated with many kinds of ecological systems each adapted to different environmental conditions and each with adaptive mechanisms that maintain continuing patterns. These are our life-support systems essential to man's survival. Many of these systems have the same kinds of occupational complexity as our industrial society, but unlike our human situation of impending environmental crisis, the ancient ecological systems continue on quietly, efficiently, cleanly self-controlled year after year even with some stress. In this chapter we examined the structure and functions of our life support and their bases in energy flows. Some hard realities emerged about potentials, yields, and the checks and balances of nature.

Once we understand the purposeful mechanisms built by natural selection into the program's control within the ecosystem, we can recognize the splendid miniaturization and complexity, which many misinterpreted earlier as a symptom of accident, disorder, and randomness. Guiding the self-managing systems of nature now seems far more sensible than the destruction of our life-support bases or dangerous, clumsy attempts to substitute our untried and expensive human anthropomechanic devices.

Many similarities were noted between complex ecosystems and our cities. Can man develop continuing patterns with nature which are stable, efficient enough to be competitive, and consistent with his present visions of himself? Can we develop man after nature instead of nature after man?

REFERENCES

[1] Bates, M., *The Forest and the Sea*, Random House, New York, 1960.

[2] Birdsell, J. B., "Some Environmental and Cultural Factors Influencing the Structuring of Australian Aboriginal Populations," in J. B. Bresler (ed.), *Human Ecology*, Addison-Wesley Publ. Co., Reading, Massachusetts, 1961, pp. 51–90.

[3] Brody, S., *Bioenergetics and Growth*, Reinhold Publishing Corp., New York, 1945.

[4] Browning, T. O., *Animal Populations*, Harper and Row, New York, 1963.

[5] Clarke, G. L., "Dynamics of Production in a Marine Area," *Ecol. Monogr.*, 16, 321–335 (1946).

[6] Conover, R. J., "Food Relations and Nutrition of Zooplankton," *Proc. Symposium on Experimental Marine Ecology*, Occasional Publ., No. 2, Graduate School of Oceanography, 1964.

[7] Darnell, R. M., "Food Habits of Fishes and Larger Invertebrates of Lake Pontchartrain, Louisiana, an Estuarine Community," *Publ. Inst. Mar. Sci.* (University of Texas), 5, 353–416 (1958).

[8] Darnell, R. M., "Trophic Spectrum of an Estuarine Community Based on Studies of Lake Pontchartrain, Louisiana," *Ecology*, 42, 553–568 (1961).

[9] Drake, J. F., J. L. Jost, A. G. Frederickson, and J. M. Tsuchiya, "The Food Chain," in *Bioregenerative Systems*, J. F. Saunders (ed.),National Aeronautics and Space Administration (NASA) SP-165, 1966, pp. 87–95.

[10] Hagen, E. E., "Man and the Tropical Environment," *Symposium on Biota of the Amazon*, 1966, Mimeo, 4 pp.

[11] Helfrich, P., and S. J. Townsley, "Influence of the Sea," in *Man's Place in the Island Ecosystem*, F. R. Fosberg (ed.), pp. 39–56, *10th Pacific Science Congress*, Bishop Museum Press, Honolulu, 1965.

[12] Jones, R. S., W. B. Ogletree, John H. Thompson, and Wm. Flenniken, "Helicopter Borne Purse Net for Population Sampling of Shallow Marine Bays," *Publ. Inst. Marine Science*, Univ. of Texas, 9, 2–6.

[13] Lotka, A. J., *Elements of Physical Biology*, Williams and Wilkins, Baltimore, Md., 1925.

[14] Ludwig, H. F., W. J. Oswald, H. B. Gotaas, and V. Lynch, "Algae Symbiosis in Oxidation Ponds. Growth Characteristics of *Euglena gracilis* Culture in Sewage," *Sewage and Industrial Wastes*, 23, 1337–1355.

[15] Mann, K. H., "Energy Transformations by a Population of Fish in the River Thames," *J. Anim. Ecol.*, 34, 253–375 (1965).

[16] Nixon, S. 1969. Characteristics of Some Hypersaline Ecosystems, Ph.D. Dissertation, Botany, Univ. of N.C.

[17] Odum, H. T., and C. M. Hoskin, "Comparative Studies on the Metabolism of Marine Waters," *Publ. Inst. Mar. Sci.* (University, Texas), 5, 16–46 (1957).

[18] Odum, H. T., "Trophic Structure and Productivity of Silver Springs, Florida," *Ecol. Monographs*, 27, 55–112 (1955).

[19] Odum, H T., "Primary Production in Flowing Waters," *Lmnolog. and Oceanogr.*, 1, 102–117 (1956).

[20] Odum, E. P., *Fundamentals of Ecology*, 2nd ed., Saunders, Philadelphia, 1957.

[21] Odum, H. T., "Analysis of Diurnal Oxygen Curves for the Assay of Reaeration

Rates and Metabolism in Polluted Marine Bays," *Waste Disposal in the Marine Environment*, E. A. Pearson, Pergamon Press, New York, 1960, pp. 547–555.

[22] Odum, H. T., R. Cuzon, R. J. Beyers, and C. Allbaugh, "Diurnal Metabolism, Total Phosphorus, Ohle Anomaly, and Zooplankton Diversity of Abnormal Marine Ecosystems of Texas," *Publ. Inst. Mar. Sci.*, 9, 404–453 (1963).

[23] Odum, H. T., "Biological Circuits and the Marine Systems of Texas," in T. A. Olson and F. J. Burgess (eds.), *Pollution and Marine Ecology, Proc. Conference on the Status of Knowledge, Critical Research Needs, and Potential Research Facilities Relating to Ecology and Pollution Problems in the Marine Environment*, Interscience Publishers, New York, 1967, pp. 99–157.

[24] Odum, H. T., "Energetics of World Food Production," in *The World Food Problem*, Vol. 3, Report of President's Science Advisory Committee. Panel on World Food Supply, White House, Washington, D.C., 1967, pp. 55–94.

[25] Odum, H. T., and E. P. Odum, "Energy in Ecological Systems," Chap. 3 in *Fundamentals of Ecology* by E. P. Odum, Saunders, 1959.

[26] Odum, H. T., and Robert Pigeon (eds.), *A Tropical Rain Forest*, AEC Division of Technical Information, Oak Ridge, in press.

[27] Patten, B. C., "Community Organization and Energy Relationships in Plankton," Oak Ridge National Laboratory Reports in Biology and Medicine ORNL-3634, 1965, 60 pp.

[28] Pearsall, W. H., "Growth and Production," *Advancement of Science*, British Association, 11, 232–241 (1954).

[29] Petrides, G. A., and W. G. Swank, "Estimating the Productivity and Energy Relations of an African Elephant Population," *Proc. Ninth International Grasslands Congr., Sao Paulo* 831–842 (1966).

[30] Ryther, J. H., "Geographic Variations in Productivity," in M. M. Hill, *The Sea*, Vol. 2., Interscience, New York, 1963, pp. 347–380.

[31] Rosenzweig, M. L., "Net Primary Productivity of Terrestrial Communities Prediction from Climatological Data," *American Naturalist*, 102, 67–74 (1968).

[32] Sanders, H. L., "Marine Benthic Diversity," *American Naturalist*, 102, 243–282 (1968).

[33] Steeman-Nielsen, E. S., "The Chlorophyll Content and the Light Utilization in Communities of Plankton Algae and Terrestrial Higher Plants," *Physiologia Plantarum*, 10, 1009–1021 (1957).

[34] Tamiya, H., "Mass Culture of Algae," *Annual Rev. Plant Physiology*, 8, 309–334 (1957).

[35] Turnbull, C. M., "The Lesson of the Pygmies," *Sci. Amer*, 208 (No. 1), 28–37 (1963).

[36] Utida, S., "Cyclic Fluctuations of Population Density Intrinsic to the Host Parasite System," *Ecology*, 38, 442–448 (1957).

[37] Vouté, A. D., "Harmonious Control of Forest Insects," in *International Review of Forestry Research*, Vol. 1, Academic Press, New York, 1964, pp. 325–383.

[38] Westlake, D. F., "Comparisons of Plant Productivity," *Biol. Reviews*, 38, 385–425 (1963).

[39] Yount, J. L., "Factors That Control Species Numbers in Silver Springs, Florida," *Limnology and Oceanogr.*, 1, 286–295 (1956).

[40] Zeuthen, K. E., "Oxygen Uptake and Body Size in Organisms," *Quart. Rev. Biol.*, 28, 1–12 (1953).

4

POWER BASIS FOR MAN

The emergence of man, the shift in his role from minor component of natural systems to predominant and sometimes exclusive occupant of modern industrial cultures, is a story of change in his basis of power support. Progress has become so rapid and dramatic in the twentieth century that many citizens believe anything is possible. That all these changes follow within the tight limits of power availabilities may not be adequately understood, for the correct leading questions about energetics are not asked in public forum, in legislative bodies, or in the general science education of the schools. Let us consider the power support systems for man in several principal categories of historical development.

MAN WITHIN A STABLE ECOLOGICAL SYSTEM

When there are no special energy sources for an ecological system except the input of the sun, man is a small part of the overall scheme, but he is protected by the great stability, complexity, and staying power of the natural system. For example, the pygmy population [33, 34] within the complex rain forest draws only a small volume from the many channels of fruit and animal products available (Fig. 3–11). The forest does the concentrating of energy and provides the upgrading of biochemical diversity needed for the nutrition of man. Similar small populations are found in the Amazon basin and in New Guinea where the soil structures and geochemical cycles are regulated and steady.

The coral atoll and island peoples of the open Pacific are other populations of man living within a stable solar-energy-based system. Aspects of an atoll system described by Alkire [2] are given in Figure 4–1. Stable

Figure 4–1 Example of low-density man in a stable system, the Pacific coral atoll at Lamotrek. The diagram summarizes in system form the data given by Alkire [2].

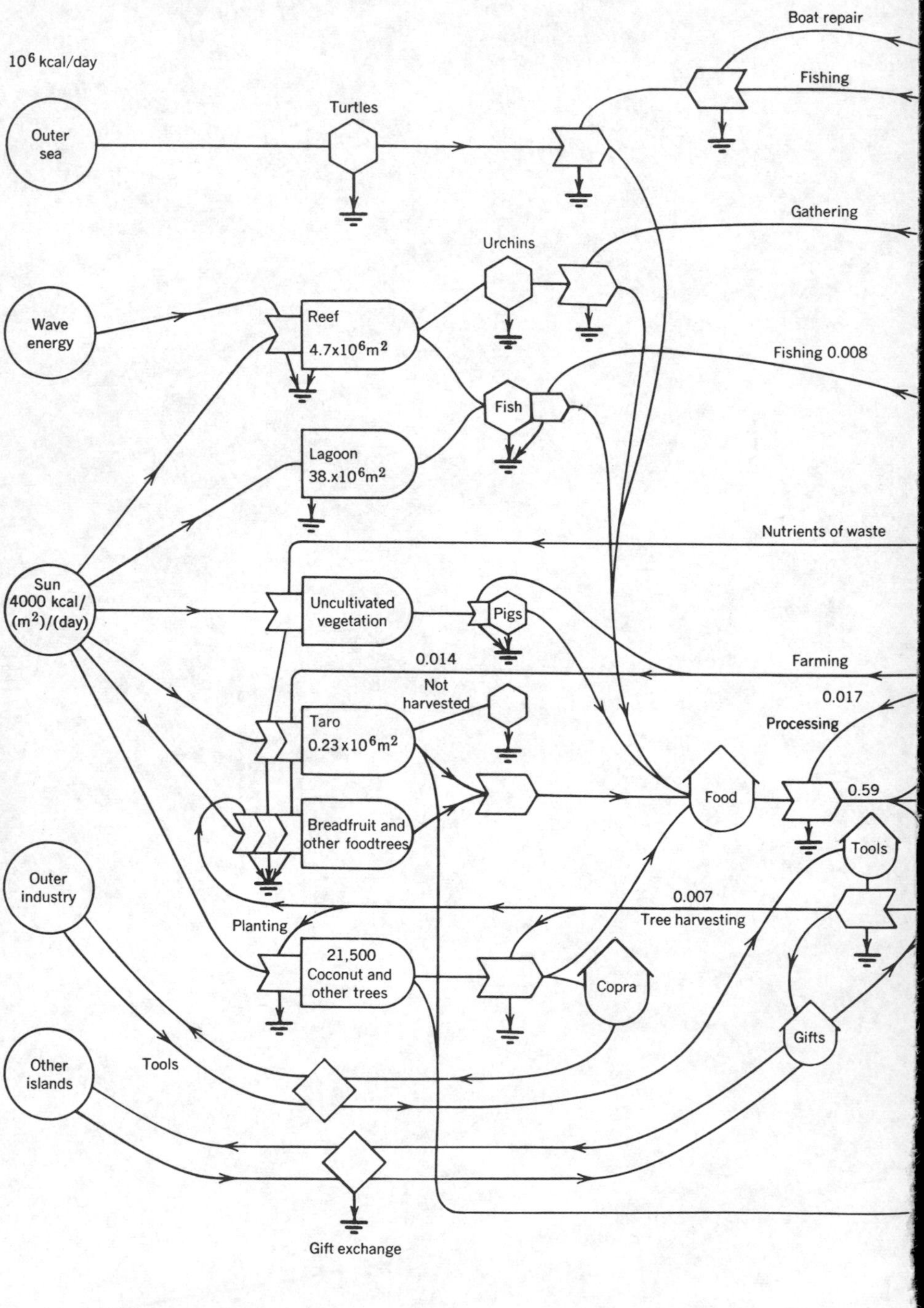

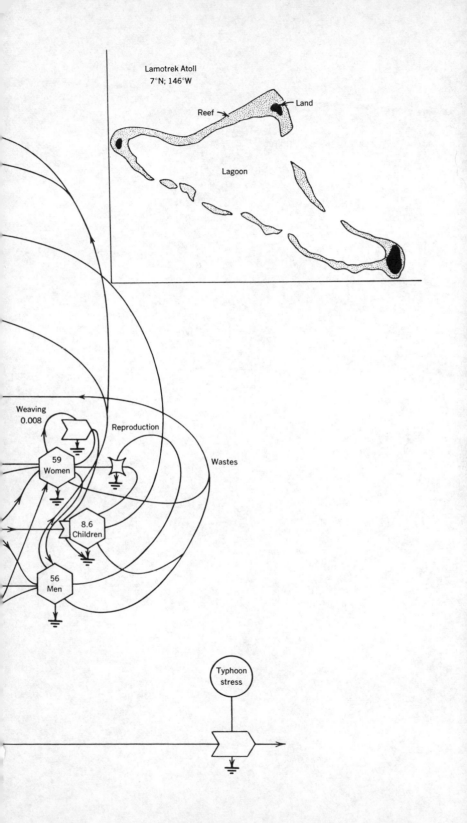

Lamotrek Atoll
7°N; 146°W

Reef

Land

Lagoon

Weaving
0.008

Reproduction

Wastes

59
Women

8.6
Children

56
Men

Typhoon
stress

climatic and hydroclimatic regimes allow for a complex network of tree products, root crops, and colorful marine inhabitants which provide many channels of food for the human population, although no great energy volume is available in any one. Staple root crops, coconuts, and bread-fruit from land support a population of one per acre but the protein source is derived from a much larger area base of reef and lagoon. Human density calculated on the whole atoll as the basis for support is 13 per square mile. The percentage of the marine energy budget for man is small, but the meagerness of man's role within a very stable system is also a protection. The energy resources available to man are insufficient for him to do damage to his supporting system. Their diversity is a protection against epidemic decimation of his food circuits.

Figures 3–10, 3–11, and 4–1 are power diagrams for man in this type of system. Since man has little energy available to direct at his system, he has little ability to increase his own role and his density remains low. His culture in such a system must include a great knowledge of species properties, of the seasonal cycles, and of the network in which he is embedded. Medicinal herbs, poisons, building materials, and animal products are available in great biochemical diversity but in small quantities so that considerable gathering energies are required. Adapting a culture to such diverse permutations of input may have been instrumental in developing the flexibility of human intelligence.

PULSATING AGRICULTURE SYSTEMS

The story for the zones of the earth where there are sharp variations and severe fluctuations in climate and hydroclimate is quite different. Where drought and heavy rains alternate or where sea salinities vary drastically, any system must either put great energies into special physiological adaptations to survive the fluctuations or, when conditions are too severe, continually establish and reestablish a system that dies back between favorable periods. In this instance special adaptations for recolonization or seasonal cycles are developed. The systems of nature developed for the fluctuating environments have little energy to specialize many species for a division of labor; all the energy is required to fit a few species to such special demands as adaptation, storage, and germination.

In other words, the systems of nature that have evolved in the zones of fluctuation on the desert borders, in the mouths of rivers, and in the polar regions are systems of a few species of pulsating populations, of temporary large storages, and of channeled energies. In such regions the natural system is much like solar-energy-based agriculture. In fact, primitive agriculture was really a capture of these existing systems in which the storages are concentrated in man and his animals. Thus systems of solar-energy-based agriculture are competitive in areas of fluctuating environment but are subject to those fluctuations, incurring frequent famine and epidemic growths.

One of man's adaptations to this system was to become a nomad, using his energy storages to move with the fluctuating zones of favorable conditions. In so doing he served as one of the means of the overall system for rapid recolonization and planting. Other animals such as locusts, caribou, bison, fish, and shrimp also migrate in these zones of the earth. Tapping the routes of these animal migrations, made on their own energy storages, provided man with a special opportunity to gain concentrated fuels.

Taking advantage of pulses, agriculture based completely on pulsating growth seasons and solar energy draws a high percentage of energy into man's culture, but it has the disadvantages of short duration and fluctuations which limit primary production during the year. The fluctuations, however, whether due to natural causes or arranged by the farmer are necessary to a harvest system. They also help keep out more complex associations of animals and plants that would divert energies from the harvest. If energy subsidies are lacking, what can be done is limited by the solar power base and the limitations imposed by competing natural systems capable of displacing the agriculture. Pulses apparently favor man because of his skills in planning. Insofar as the energy budget permits, the agricultural system can ameliorate the disadvantages of the environmental pulses with special mechanisms such as microclimate control and storages. Some very effective systems of man and agriculture have been evolved in the monsoon, prairie, and savanna belts of the earth.

Rice-Cattle Systems in Climates with Sharp Dry Seasons

One of the most successful ancient systems for supporting dense populations based wholly on solar energy in monsoon climates is the small rice plot in which farm animals are used as energy transformers to do the necessary work of planting, fertilizer spreading, high-protein synthesis, energy storage, and weeding. The energy circuit for the rice-cattle system is given in Figure 4–2. Man and his animals have the energy storage to inject a fast start for the agricultural crop, putting it ahead of possible competitors when the cool moist winds from the Indian Ocean return in June and allow both man and nature to start over after the parching 120-degree spring. Similar systems are applied to a lesser extent in other tropical areas where there is a sharp climate or other artificially induced pulses.

RELIGION AS AN ENERGY PROGRAM: SACRED COWS

Throughout much of Asia we find variants of this system—man living in high density of one per acre with his farm animals and intensely cultivating a crop such as rice. The energy flows, as estimated in Figure 4–2 for small plots in India, include the work roles of man and his farm animals and their other functions such as nutrient cycling and processing dung for fuel and fertilizer. In India the cows are sacred to part of the population, and many scientists have advocated their elimination in order to skip the extra mouth and shorten the food chain. An examination of the energy flow diagram, however, suggests a different judgment, for the network runs well on light energy with control loops and features of stability in the otherwise unstable environment of the monsoon. Perhaps

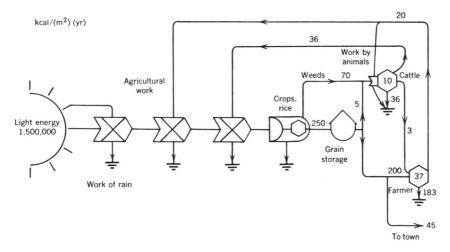

Figure 4–2 Man in an unsubsidized agricultural system in a country such as India whose climate has a sharp seasonal pulse. Relationships were suggested by Harris [14]. Data are for tropical dense populations with 640 persons per square mile, 0.1 animal per person. Indian grain yields 250 kg/(acre)(yr) [4]. One farm animal is shown for each ten persons. One-third of the food calories of cattle remain in feces. Work and fecal fertilization are taken as half of animal metabolism. The animal protein intake for India is about 6 g/person [4]; 2 percent of the food crop is fed to animals. Animal metabolism is 8000 kcal/day [18]. Farm work is taken as 0.1 of total man-hours of population.

Harris [14] is right in opposing those who wish to eliminate the sacred cow. He says they glean from different food chains, produce the bullocks that are necessary to the agricultural pulse, recycle the minerals, and provide some critical proteins through milk. He quotes Ghandi's statements that the cows are sacred because they have practical value.

Perhaps there is a more general principle that religion has practical value as a program of control for energy flows. As long as the energy bases are the same, the old religions are a pertinent and automatic means for controlling the established system for survival. When new energy sources emerge, perhaps the religious programs must be revised, as in our day of new fossil fuels.

A NATIVE CATTLE-KEEPING SYSTEM

Figure 4–3 is a network diagram with mean flow rates for an African tribe subsisting on the dual flows of crop and cattle products. The system has only 11 percent of the land under cultivation, and the cattle drain only 18 percent of the net plant production. Thus, the natural ranges with their adjustment based on a natural diversity of animals, plants, and microorganisms are prevalent, embedding man in a stable system and supplying him with all crop yields. Any replacement using more of the natural ranges would require the substitution of a control system at greater energy cost, raise the prospect of overgrazing the photosynthetic surfaces, and upset the soil bases.

The photosynthetic efficiency of visible light is 0.31 percent, being water-limited and in the range for which Walter (see E. P. Odum [24]) provides yield figures related to rain. Small work expenditures by man just over 3 kcal. control and augment flows of much larger magnitude, thus serving as gates and having their own work amplified by the natural system. The yield of cattle protein to man is 0.06 percent of the range output produced by the many processes of collecting sparse vegetation, putting it through ruminant microbe systems of the stomach, and synthesizing not just ordinary organic matter but high-quality diversified proteins while at the same time refertilizing the range to stimulate the system further.

THE SEMIDOMESTICATION OF TROPICAL RUMINANTS

In the tropical zones that are subject to sharp seasons, such as the monsoon belts and the savanna belts, the stronger pulse of nature provides for greater temporary accumulations in the same manner that large fish

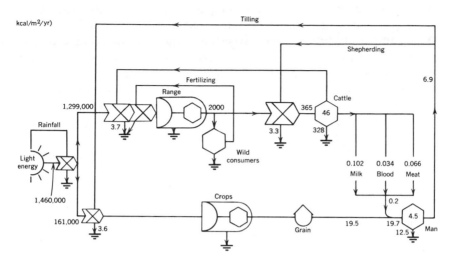

Figure 4–3 Example of a simple agricultural system in a pulse climate, the Dodo tribe in Uganda (based on data and account given in Ref. [9]). Food is derived from grains, meat, blood, and milk. Animals serve as a storage filter, smoothing the pulse, and as a nutritional convergence point. Measurements are given in kcal/(m²)(yr). There are 70 people per square mile.

Dry weights are converted to kilocalories using 4.5 kcal/g. The basic net production for dry regions with limited rainfall is given by Walter (1954), reproduced in E. P. Odum, [24, p. 403] as a function of rainfall. Using 21 inches of rainfall and Walter's diagram we obtain 500 g dry matter/(m²)(yr) for the net plant production. With one acre cultivated per person and 70 persons per square mile, 11 percent of the natural yield area is preempted by crops.

At 4.5 cattle per person, 560 pounds per cow, and 33 percent dry weight excluding ash and water, we find the stock to be 10.2 g/m².

If we integrate the area under the curves given for monthly consumption of milk, blood, and meat, annual calorie yields per man from the cattle are 3800 kcal milk, 2450 kcal meat, and 1265 kcal blood, these yields providing the per area data. The calorie requirement per person is given as 2000 kcal/(person)(day) or 19.7 kcal/(m²)(yr). The milk, meat, and blood supplies only 0.2 kcal/m²/yr, so that the crop intake is 19.5, a net yield much less than that of vegetation of the natural range.

Total solar insulation in this area just above the equator is about 4000 kcal/(cm²)(day) based on Drummond's (1958) solar radiation maps for winter and summer in Africa [36].

The work of man in tending the crops and cattle can be taken as the percent of his time involved in this activity (primarily the daylight hours). In a culture intimately involved with cattle, we may attribute one-sixth of the daily metabolism of man to tending cattle and the same to tending crops, assuming that the various aspects of the maintenance of man during his work are necessary to that work. The metabolism of 650-pound steers estimated from Kleiber [18] is 8000 kcal/day or 365 kcal/(m²)(yr).

Some fraction of the cow's time and metabolism goes into refertilizing the range

and game stocks developed in the temperate latitudes, permitting man to develop nomadic and hunting cultures based on an animal food system. A number of authors have suggested harnessing the natural animal herds of Africa as preferable to the importation of traditional domestic herds. Especially where full energies for protection and intensified care are lacking, a moderately diversified range based on several game animals might provide a yield of intermediate wealth and ultimately lead to an intensified agriculture with genetic modifications of these species. The stocks of the African range carry the microbial symbionts that are necessary to convert plant yields into the nutritious mixtures from which meat is developed (see Fig. 3–13(a) for details of microbes in ruminants).

Figure 4–4 shows the network for an African savanna with a few of the flows and stocks enumerated after Petrides and Swank [29]. The system receives loop stimulus with several work flows drawn from the downstream consumers to the production system. Some vegetation goes into fire, thus making the edible, less woody vegetation more prevalent; the fecal-microbe-invertebrate system regenerates the minerals to the plants in the form and ratio suitable to reinforce the species favored by the animals involved, thus closing the loop of mutual reward. Theoretically, man and other carnivores may draw stocks from the large animals if through their own activities and work they feed back a stimulus to reward the desirable system to the extent of the meat drain, thus preventing the invasion of a competitive system not harvestable by man. Such rewards include stock selection improvement, fertilization, and the limiting of stocks during periods of excess population. Such a loopback may always have existed in the relation of native peoples to the savanna. Removal of the tribal hunting is producing overgrazing problems. In some overgrazed African big game parks, shrub invasions have been accompanied by proportionate increases in elephants. Perhaps they have a role in cutting back potential invaders of their basic system by browse preference programming. Harvesting the large animals may remove some of the system management. If a modern system involves some fossil fuel, this fuel may support enough wildlife management and stimulus of the elephant grass and related plants to compensate adequately for the downstream harvest of animals which were managing the range.

---

on which it grazes, thus reinforcing its loop and maintaining its competitive position. Part of a cow's day is spent on the move, and some of its organ systems are involved in the nutrient regeneration system. One-tenth of its metabolism is taken as its work contribution to stimulating the vegetation.

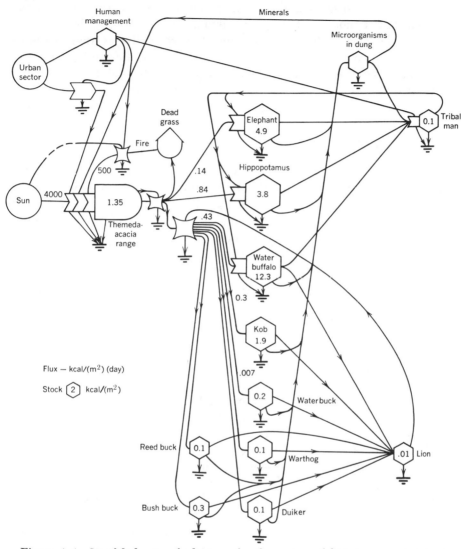

Figure 4–4 Simplified network diagram for the system of big game ruminants on African savannas. Data are from Petrides and Swank [29] from a moderately grazed area of Queen Elizabeth National Park in Uganda.

As long as management does not increase the animal density the savanna has the diversity for stability protection, a fairly high biomass stock, enough storage to deal with the pulses of drought, and a means for concentrating stocks at water holes when there is little growth and expanding them again to the plains as rainfall develops grass. It is difficult to imagine how minor injections of fossil fuel might improve the productivity of the meat network or manage it more easily for that climate. Because the savanna self-managed in this way is a meat system, its carrying capacity for man may not be high, but the stability and standard of nutrition may make the system preferable to a wildly oscillating agricultural system.

SOLAR AGRICULTURE VERSUS COMPLEX FOREST

A later stage in the evolution of systems of man and nature, possibly paralleling in time early agricultural evolution, involved attempts to introduce agriculture into a stable forest system such as the rain forest. Whereas grain agriculture in prairie and savanna climates was much like the seasonal restart of grassy natural systems, agriculture was quite different when complex forests had to be displaced.

Every schoolboy learns that historically man began his agricultural role through domestication of plants and farm animals. In such a system the domesticated plants and animals are substituted for the more complex and diverse network of the stable system. As diagramed in Figure 4–2, the channeling of energy and the insertion of much of man's power drain at an earlier stage of energy transformation theoretically provides more energy to him and ultimately more activity and carrying capacity for his culture. Yet such a simplified system of fewer species has lost features of the stable type: the biochemical diversity, the many diverse control circuits, the protection against epidemic decimation of his essential populations, and the means for soil maintenance and weed control.

Now compare the overall power processing of the two systems, both on the same input of solar power. As long as the environment is stable, the diverse type gradually displaces the simplified version because the diverse one has staying powers and self-developing mechanisms that the simplified version lacks. The simplified system has diverted energy away from stabilization of geochemical cycles and soils, from epidemic control, and from all the many special adaptations of the complex system. It is a hard fact that the diverse ecological systems successfully compete with and displace the simplified systems when environmental conditions are stable and there are no periods of severe cold, drought, or other life-limiting circumstances. Unless special energies are directed to prevent succession, the complex system with its specializations performs more work functions for a total effort and displaces systems whose energies are going into storage instead of into useful work for competitive survival and dominance. It is a basic ecological principle that succession starts with an early colonization of a few species specialized for accumulation, for in the next season different and diverse combinations of species use the minerals and materials produced by the earlier fewer species, cover the ground and displace them, and develop organization and work functions. Diversity, stability, and order result.

ARTIFICIAL PULSE AND SHIFTING AGRICULTURE

If, however, some of the energy of the system is recycled into converting it to a pulsing one with continual restart (Fig. 4–5), agriculture becomes possible at least during the start-up periods. The energy accumulated in succession may be used to undo its own work. The farmer may cut the growth and burn the cuttings in order to restore the system to a starting point again. Such procedures have evolved the world over, and under population pressure they are invading many stable climate zones of the primitive-man-in-forest type. These procedures are called shifting agriculture.

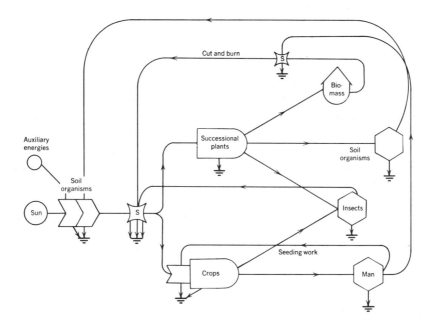

Figure 4–5 Energy diagram for shifting agriculture. Man serves as a switching timer, transferring the flow of energy into food crops made temporarily possible by the material and work accumulated during the long period of plant succession between short agricultural cycles.

One of the world's great problems is the migratory pattern of slash and burn agriculture. Tropical forests are cut, the ground is tilled for one to three years, and then the area is abandoned before the complex, diverse system forces its succession, displacing the farmer. The farmer's plot, if it were not abandoned, would not hold the soil structure, nor provide him enough energy to control the successional invaders (weeds and insects), nor hold stabilized chemical cycles. As long as the farmer has available only the same energy resources of the sun that the competing complex system has, he must rotate sites for lack of power. Carter [6] details the labor and social structure that accompanies a workable shifting agriculture in Guatemala.

Man can live in the slash plot for a year or two because he has taken advantage of an energy storage of the forest for a quick and destructive drain comparable to a short circuit. He is able to use the potential energy of the elevated crowns amassed over many years to remove the competing vegetation by force of gravity, and the accumulated conditions in the soil will produce a crop or two for him. Since the natural system has as one of its mechanisms a whole range of species for quick reforestation of openings, the diversity of the forest is focused to defeat the competing system rapidly. The forest soil has a storage of seeds which explode in germination when released by the action of the sun and by soil disturbance.

The phenomenon of millions of acres being used for such a brief percent of the time is a necessary outcome of the energetics of the competitive diversified systems produced by stable environmental conditions. On a long-range basis man's yield from the forest with a year or two of a solar-energy-based agriculture every 20 to 50 years may be of the same order of magnitude as his living within the forest system in stable tribal adjustment.

Man obtains net yields from pulsing systems, which in this case he makes by his own shifting pattern of cutting. A report prepared by the President's Science Advisory Committee [30] has described a modern tropical system of agriculture in the Congo which uses corridors of rotation and is based on an old system of shifting cultivation developed by the Bantu peoples. The short cultivation phase ends with a shade crop which helps reestablish the complex successional vegetation quickly, thus restoring the mineral cycle, soil structure, and the canopy that will permit a new cycle after later cutting.

The network diagram for shifting agriculture in Figure 4–5 has two main energy flows which switch first from one to the other, with man controlling the action through a small expenditure of cutting energy.

The timing is ultimately determined by the period required to build up the energy storages that permit an effective fall and burn, these functions serving to clear competitors, eliminate insects and seeds, regenerate minerals, and otherwise set up a short period of crop yield. Systems engineers will recognize the automatic flip-flop circuit. Man can have continuous service by operating himself in rotation between many plots —the pulses of each on a different schedule of cut and regrowth.

MAN WITH FOSSIL-FUEL SUBSIDY

Beginning in the last century, man began to develop an entirely new basis for power with the use of coal, oil, and other stored-energy sources to supplement solar energy. Concentrated inputs of power whose accumulation had been the work of billions of acres of solar energy, became available for manipulation by man. Knowledge was required to develop machines capable of coupling these power inputs to man's needs, but the real force causing progress was the entirely new order of magnitude of the energy sources. There were excess energies for further research to obtain more fossil fuels and involve them in driving cultures. The subsequent progress was like a flash explosion compared to the steady fires of the long evolutionary record for previous millions of years. Thoughtful leaders warned of the things to come when stored fuels were gone, but then nuclear energy arrived.

POTATOES PARTLY MADE OF OIL

One of the results of industrialization based on the new concentrated energy sources was abundant food rolling out from huge fields which were sowed with machinery, tilled with tractors, and weeded and poisoned with chemicals. Epidemic diseases were kept in check by great teams of scientists in distant experiment stations developing new and changing varieties to stay ahead of the evolution of disease adaptation. Soon a few people were supporting many, and most of the rural population left the little farms to fill the new industrial cities.

A very cruel illusion was generated because the citizen, his teachers, and his leaders did not understand the energetics involved and the various means by which the energies entering a complex system are fed back as subsidies indirectly into all parts of the network. The great conceit of industrial man imagined that his progress in agricultural yields was due to new know-how in the use of the sun. A whole generation of citizens thought that the carrying capacity of the earth was proportional

Table 4–1   Agricultural Food Production

Crop	Harvest of Organic Matter kcal/(m²)(day)	Efficiency in Use of Sunlight (%)[a]
Not subsidized by fossil fuels		
Farms, United States, 1880[b]	1.28	0.03
Grain, Africa, 1936[c]	0.72	0.02
Industrialized agriculture		
Rice, United States, 1964[d]	10	0.25
Grain average, North America, 1960[b]	5	0.12

[a] Calculated with 4000 kcal/(m²)(day) sunlight.
[b] Parker [27].
[c] Brown [4].
[d] President's Science Advisory Committee on World Food [30].

to the amount of land under cultivation and that higher efficiencies in using the energy of the sun had arrived. This is a sad hoax, for industrial man no longer eats potatoes made from solar energy; now he eats potatoes partly made of oil.

Table 4–1 lists some data on production and energetic efficiency of converting solar energy by agricultural systems. Compare these with data on production by the natural systems developed through millions of years (Table 3–3). The systems man has developed have not increased the initial conversion of the energy of the sun at all (except locally where some raw material was lacking and added by man). In fact, the gross photosynthesis (P) of the natural systems in each belt of the earth provides the upper possible limit of the sun's conversion for any alternative system without subsidy. Usually the conversion in man's system is much less because the requirements of weeding and harvest leave the ground poorly covered with chlorophyll or bare a greater part of the year. If the gross photosynthetic conversion of the solar energy has not been improved, what then is responsible for the great improvement in acre yields of agricultural or algal products?

In Chapter 3 we learned that most of the energy of the natural ecological cover had to be cycled into the many functions of staying, competing, recycling, and holding structure. If all these functions are now carried out by a new lateral input of energy, the gross photosynthesis formerly diverted is now available for storage and yield to man. We may state the fact another way. When there are implementing pathways, the total energies available to the system are the sum of the total power flows of concentrated potential energy. The first priority goes for elimination of limiting factors.

The energy diagrams in Figure 4–6 show these relations. The fossil-fuel energies are put into the circuit at the point where there was feedback work and in a magnitude greater than that derived from the solar energy concentration process. To compete with the alternative natural systems that could invade, we need only put more energy flow into the protection of the desired system. As seen in Figure 4–6(a) the potato reaching man is produced partly by the energy subsidies of the culture. Shown in Figure 4–6(b) are diagrams comparing animal husbandry with and without the fossil fuel work subsidies.

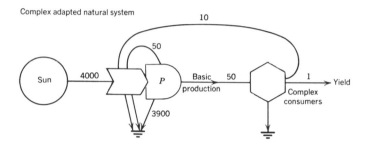

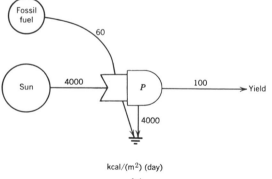

kcal/(m²) (day)

(a)

Figure 4–6 Comparison of a complex natural system adapted to maximize its basic production through the works of its diversified organization of consumer species with the same system after fossil-fuel supported works of man have eliminated the natural species and substituted industrial services for the services of those natural species, releasing the same basic production to yield. (a) Plant production and yield; (b) animal production and yield.

Much of the power flow that supports the agriculture is not spent on the farm but is spent in the cities to manufacture chemicals, build tractors, develop varieties, make fertilizers, and provide input and output marketing systems which in turn maintain mobs of administrators and clerks who hold the system together. As we stand on the edge of the vast fields of grain with tractors and production as far as the eye can see, we are tempted to think man has mastered nature, but the plain truth is that he is overcoming bottlenecks and providing subsidy from fossil fuel. Figure 4–7 shows the overall energy input to a United States farmer. Wherever the flow of industrially organized fossil fuel is missing or the work from a fossil-fuel culture eliminated, the agriculture possible is exactly what it once was or worse, for the know-how of primitive systems and the necessary varieties may not be available.

Self-maintenance on range

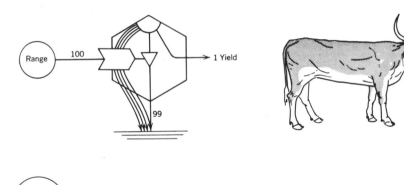

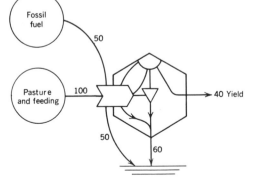

kcal/(m²) (day)

(b)

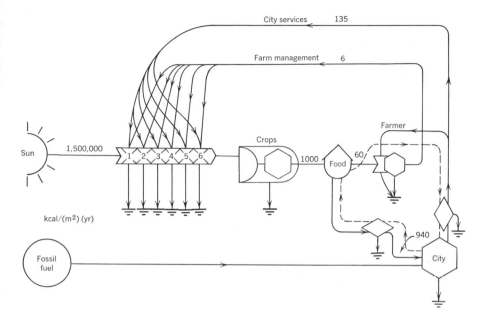

Figure 4–7 Man in a system of industrialized, high-yield agriculture. Energetic inputs include flows of fossil fuels which replace the work formerly done by man, his animals, and the network of animals and plants in which he was formerly nursed. Work flows include the following: (1) mechanized and commercial preparation of seeding and planting, replacing the natural dispersion system; (2) fertilizer excesses which replace the mineral recycling system; (3) chemical and power weeding, replacing the woody maintenance of a shading system; (4) soil preparation and treatment to replace the forest soil-building processes; (5) insecticides which replace the system of chemical diversity and carnivores for preventing epidemic grazing and disease; (6) development of varieties which are capable of passing on the savings in work to net food storages. New varieties are developed as diseases appear, thus providing the genetic selection formerly arranged by the forest evolution and selection system. In this system 170 persons per square mile support 32 times this number in cities. The level of grain production in the United States is about 1000 kcal/(m²)(yr) [4]. The fuel subsidy is calculated using 10^4 kcal/dollar. If production yields $60/(acre)(yr) and if the costs are 90 percent of the gross, then $54/acre is the measure of use of materials and services from the industrialized culture. This becomes 54 × 10^4 kcal/acre or 135 kcal/(m²)(yr).

What a sad joke that a man from an industrial-agricultural region goes to an underdeveloped country to advise on improving agriculture. The only possible advice he is capable of giving from his experience is to tell the underdeveloped country to tap the nearest industrialized culture and set up another zone of fossil-fuel agriculture. As long as that country does not have the industrial fuel input, the advice should come the other way.

The citizen in the industrialized country thinks he can look down upon the system of man, animals, and subsistence agriculture that provides some living from an acre or two in India when the monsoon rains are favorable. Yet if fossil and nuclear fuels were cut off, we would have to recruit farmers from India and other underdeveloped countries to show the now affluent citizens how to survive on the land while the population was being reduced a hundredfold to make it possible.

Figure 1–2, which shows an island economy before and after the new fuel input, summarizes the patterns of change in converting from a solar-energy-based to a fossil-fuel-based agriculture.

ENERGY SUBSIDIES FOR FOOD PRODUCTION

Measurements of the gross photosynthesis of natural systems in various situations can provide estimates of the potential food production with an outside energy subsidy. Let us consider two categories of systems, those limited by some factor such as water shortage and those without limiting factors other than the quantity of incoming light energy.

When limiting factors are a problem, the well-adapted natural eco-system that has existed for a considerable period of time has developed mechanisms to overcome these factors within its budget of available energy. Feedbacks of concentrated energy in the form of work done by higher organisms serve to eliminate the limits. For example, desert plants use parts of their energy to support water-storing adaptations, deep roots, and means for opening stomata only at night. If, however, some limiting factors are still beyond the resources remaining to the natural system, relatively small subsidies of fossil-fuel energy can be directed to eliminate them. The added energy, by increasing the flow through the limiting gates, is amplified and augmented in effect beyond the contribution it might make as a straight substitution elsewhere in the system. Without the outside subsidy, the gross photosynthesis of the natural system probably cannot be excelled, for it is using all its yield to solve limiting problems. The food production system that man substitutes has the additional load of supporting people.

In those areas of favorable growth with *few limiting factors,* the natural systems may have already overcome all obstacles except the inherent one of limited light. The situation is diagramed in Figure 4–6(*a*). The potential food production is the gross photosynthesis of the adapted natural system. Even though natural complexes export little, they do have high gross photosynthesis (Table 3–3). For such systems the fossil fuels cannot overcome limits, but they can be substituted for the energies being looped back into work processes. As more and more calories from outside are substituted for necessary work, the net food yields may be increased toward the gross production. By substituting for system work, oil is energetically converted to food up to the extent of gross production before subsidy. The system after substitution is shown in the lower diagram of Figure 4–6 (a); yields attained over longer periods of time are given in Table 4–1.

WORK OF ACCUMULATING HIGH-QUALITY
HUMAN NUTRITION

In addition to supplying an adequate total daily calorie intake of potential energy, the food of man and of other organisms of high functional complexity must be nutritious. It must have a great diversity of the necessary chemical molecules in ratios suitable for the support of man. Many systems that support man have two types of food input, one flow a staple which supplies calories in a carbohydrate form such as potatoes, the other a protein supplement which supplies the daily requirement for diversity of molecules. Figures 4–1, 4–2, 4–3, and 4–4 show these double flows. The potentials for the supply of calorie food staples are the potentials for net plant production, as we have already discussed. The high quality foods require more processing work combining variety.

Mass production either by natural systems or by man's industry requires focus and effort on one crop and one kind of yield. Unhappily the change of many systems from diversified natural ones to specialized modern mass production has diminished the relative protein production and has raised the specter of malnutrition. If the carrying capacity of the land increases for carbohydrates and fat without the protein flow also being augmented, a new limiting factor for man develops, one that was not so often as critical in the earlier systems. Species diversity is necessary for nutrition, but with diversity the advantages of mass production are lost. The variety of nature's original habitats provided man with many foods when he was a small part of the system. Later when his agriculture was concentrated on a few plants and replaced nature with a greater net yield of food, his carrying capacity increased but some of the energy converted to increased yield was at the expense of the former diversification. To supply the nutritional diversity again requires an energy expenditure for diversity, either through transportation of products from elsewhere or through local rediversification.

A man can eat adequately without meat if he goes to the extra work and hence energy expense of gathering enough kinds and diversity of plant foods. As an alternative he can eat the more expensive meat products which already provide a nutrition close to his need. In meat the combining has already been achieved by the integrating and mixing actions of the animals, microbes, and their systems of work and sustenance (Fig. 3–13(a)). Other alternatives are fossil-fuel supplements to enrich the nutrition with ingredients for particular needs, again using the rich supplement of concentrated fuel energies processed through new chemical and microbiological means.

As we have seen in our discussions of the complex natural rain forest system (Figs. 3–11 and 3–12), chemical diversity is built up in food through the successive steps of food chains so that carnivores eaten by man are close to the nutritional need of man. For example, in the Peruvian Amazon, crop harvests are supplemented with some wild game as a satisfactory means of providing protein. The various chemical molecules are synthesized by specialized plants and microorganisms in diverse initial pathways and are then combined and selected in the convergence to man. The energy cost of this convergence is large because something on the order of 10 percent or less of the potential energy eaten at one step is eaten at the next. We must ask whether it can be done more cheaply.

Any one plant can be produced with high yield, but even within a single species selection for high yield tends to sacrifice diversity. Small seeds without much accompanying nutrition are produced in large quantities by successional trees, but the large fruits in mature forests can only be produced in smaller numbers. Diverse nutrition requires that considerable energy be processed into the work of maintaining, controlling, and coordinating dissimilar manufacturing systems, many enzymes, and many more aspects of genetic inheritance. Yields of carbohydrates are thus cheaper in work energy spent than are the high-quality foods.

Figure 4–8, an energetic diagram, suggests that the energy costs for chemical diversity follow the same principles that require large energy costs for species diversity, a related phenomenon (Figs. 3–9 and 3–11). If this reasoning is correct, high-quality nutrition for man must have a considerably larger energy base than is implied by his caloric needs alone. Either we use a complex natural system of diverse plants and microorganisms converging its products through the energy-expensive food chain system, or we set up specialized manufacturing processes for each needed entity and through a large organization of collection and distribution provide this nutrition. Either way it is much more expensive in energy or dollar equivalents than the single food circuit. See Fig. 3–9.

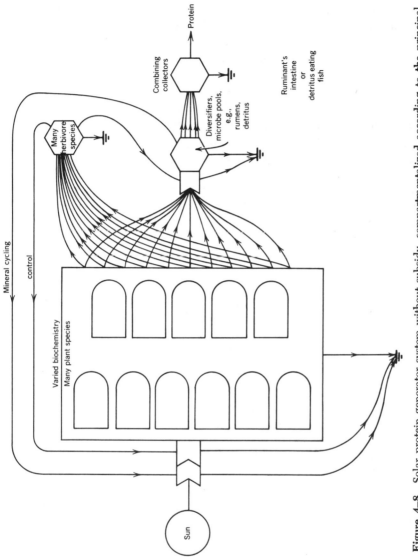

Figure 4–8 Solar protein generator system without subsidy, compartmentalized according to the principal necessary functions—producer, diversifier, controller, combining selector, and man. Each block function may involve many species and their energies for coordination. Examples are indicated.

123

MAN AS A COMPLEX NETWORK

In some ways man is like an intricate controlling device into which converge many inputs, all of which are required for the complicated function of a complex programmed unit. Within man there is great diversity and specialization, but many of the inputs require specialized molecules—the vitamins, amino acids, and minerals—and there are inputs to his behavioral and sensory systems as well.

Man is as complex a system of differentiated units as a whole community of organisms in a forest or on a reef. We speak of 3000 calories of food input required for man as his daily potential energy requirement, but special components must be present in particular qualitative proportions or the systems will develop some limiting condition which will damage the individual's entire operation. Thus the carrying capacity of an area may not be computed on the basis of gaining 3000 calories of energy, for also required are the special components, each of which has an energy cost. The energy value of a vitamin is not its potential energy value as a fuel but the calorie expenditure required to manufacture it and deliver it to man.

Thus man cannot live on a single plant product. Either he has diversity in plant foods, or he eats an animal that is close to his biochemical needs after the animal has gone about achieving the necessary diversity through its food chains.

WORLD STANDARD OF LIVING

There was a time when the world's standard of living in various places could be discussed in terms of the distribution of the solar energy, the concentrated foods provided in some places by nature, and the distribution of local concentrations of fertilizer, soil, and water. Now, however, fossil fuel (or nuclear power if its use becomes important) supports food-manufacturing systems and general culture. Population growth and crowding is diluting the many inputs that formerly came from the natural systems making us more and more dependent on the fossil fuels. Many thoughtful leaders have long warned of the limitations of fossil fuels, although the timetables of disaster are continually being revised as new sources of these fuels are found. But the earlier warnings are pertinent even if the timing of crisis is still unknown. Oil and coal will not run out, but the ratio of energy found to energy spent in obtaining them will continue to increase until costs exceed yields. If the net yield of potential energy begins to approach that of wood, we will have returned to the solar-energy-based economy, and by that time the standards of living of the world will have retrogressed to those of two centuries ago. Whether such changes will come suddenly in a catastrophe or slowly as a gradual trend is one of the great issues of our time.

A CARRYING CAPACITY FOR MAN

In wildlife management the phrase "carrying capacity" is sometimes used to describe the ability of a grassland range to sustain a level without damaging the network of supporting plants, mineral cycles, and means for maintenance of soil, water levels, diversity, and reserves that protect against fluctuations. Carrying capacity is the population level for long-range survival.

The essence of the problem of food production for the world is: "What is the carrying capacity of the earth's surface for man?" The same question arises in the discussions of man in space. On the surface of the moon, what is the area necessary to support man on solar energy? How much area is needed when solar energy is supplemented by some fuel from earth? The biosphere is really an overgrown space capsule, and the questions about carrying capacity are similar. For projected levels of energy supplement from coal, oil, and nuclear power, what is the carrying capacity of the earth? Support of man must include all the necessary works that provide stability, reserves, protection, and all the crisscross controls required to regulate his many complex needs. Either we must retain the old network of hundreds of species of plants and animals which carried on these functions for man, or we must set up new machinery to perform them on energy subsidy. The carrying capacity of one man is the minimum area into which the work to support him can be concentrated.

The only real data we have come from the underdeveloped countries before they are disturbed by being partly involved in the fuel-rich economies. The small one-acre system of a simple farmer, his animals, and his soil regeneration system is one example (Fig. 4–2). Another example is primitive man, one per square mile, in the deep stable forests (Fig. 3–11) or coral atoll (Fig. 4–1). These systems use considerable acreage to support man. They are running on solar energy alone.

ALGAL CULTURE AND FALLACIOUS DREAMS

A cruel illusion was proffered by laboratory scientists and writers who proposed that we could feed the world on algae which they implied were productive on a different order of magnitude from agriculture.

The carrying capacity for a man was said to be only half of a square meter. Two main fallacies were involved, but before they could be proved, those with a little knowledge were attracted by the subject and popularized it with articles published in leading magazines. Part of the belief that the surface of the ocean holds some special hope for feeding expanding populations comes from these sad efforts which spread misinformation. When practical attempts at farming were made, the yields came out to be the same as those of maximum agriculture and were a great deal more difficult to connect to man's needs.

One of the fallacies was the extension of the high efficiencies that are inherent at low light intensity (1/100 daylight) but are not possible for energetic reasons in regular daylight (refer again to Figure 2–2 showing power and loading). Figure 4–9, based on the same law, shows the efficiency when increasing input power such as light (I) is applied to energy-transforming machinery such as plant leaves. Systems being run close to stopping do have a high conversion efficiency but, as an appreciable yield develops, energy laws require a decline. The optimum loading for maximum one step yield is at 50 percent as shown (see Figure 11–2).

The second fallacy lay in considering the algae free from its supporting ecological system which in nature requires mineral cycles, the diversion of much energy into various networks for stability, and the maintenance of extensive work for survival, as we have already discussed. In the laboratory tests all these elements were supplied by the fossil-fuel culture through thousands of dollars spent annually on laboratory equipment and services to keep a small number of algae in net yields. Scaling this expense up naturally produced fantastic costs. Until all these dollar costs were considered, the advocates imagined that they had higher efficiencies

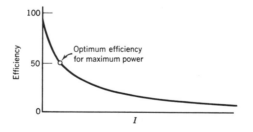

Figure 4–9 Efficiency of energy transformation when the output load is constant while the input is varied. This curve is well known in plant photosynthesis studies where light is varied and photosynthetic production is measured. It applies to any input-output energy transformation and follows from the behavior of opposing loads given in Figure 2–2.

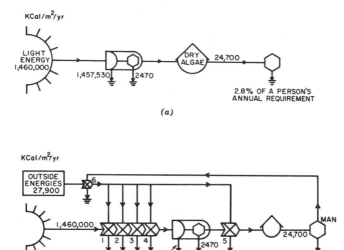

Figure 4–10   Energy flows in a food-producing algal pilot plant operated by Arthur D. Little Company [11]. (*a*) Energy costs of management omitted; (*b*) energy costs of management included. In (*a*) are the flows per square meter per year that the algal protagonists included in their calculations, finding 4.2 percent efficiency of visible light, that is, half the solar input energy is shown. Fossil-fuel subsidies are omitted.

A yield of 20 tons/acre was obtained. A year of insolation provides 1,460,000 kcal/m input (400 langleys/day). Gross production is taken as 10 percent greater because of respiration, which is small in rapidly growing and harvested cells since they are not maintaining their own system. To convert algal weight to potential energy of yield, 5 kcal/g dry weight was used.

The work costs, given by Fischer in dollars, include a depreciation of the cost of the installations over a ten-year period. The dollar costs of $2.8/(m^2)(yr) were converted to work energy values for the United States in 1960, about 10,000 kcal of fuel consumption per dollar spent. Work flows are: (1) fertilization, (2) stirring, (3) containing and distributing, (4) controlling growth and (5) concentrating for harvest.

of net yield. The erroneous calculation is given in Figure 4–10(*a*); compare this with the actual situation in Figure 4–10(*b*). Gross yields were the same as high yields in other systems, and net yields were higher because fossil fuels had provided the work necessary for system survival, in the same way that fossil fuels now subsidize potatoes. How many citizens heard the outcome of this fantastic means for putting public funds into research? If the scientists involved had been more broadly trained to see that algae are part of an ecological system, whether they are in laboratory glassware or in the sea, they would scarcely have had the nerve to proclaim their physiological researches as being of such importance to survival.

IMPROVEMENT OF FARM ANIMALS AND VARIETIES

Another of the conceits of our industrialized culture is our pride in having improved varieties of beef cows, of corn, of chickens, and of all the domestic strains that give man yields many magnitudes over those achieved in primitive societies. The citizen thinks that this is a permanent advance which secures for agriculture a high yield and that we will never have to go back to the scrub animals.

This belief is based on energetic fallacies. Genetic breeding has carefully eliminated all the special energy-using abilities of these animals so that their food can go into maximum net growth. Man supplies all the old functions with his cultural fossil-fuel subsidies.

We now have chickens that are little more than standing egg machines, cows that are mainly udders on four stalks, and plants with so few protective and survival mechanisms that they are immediately eliminated when the power-rich management of man is withdrawn. Such varieties are complementary to the industrialized agriculture and cannot be used without it. See Figure 4–6(b).

If man has agriculture without this subsidy, he must have cows that can fend for themselves in reproduction, protect themselves from weather and disease, move with the food supply, and develop their own patterns of group behavior. He must have competitive plant species that can also provide reasonable growth in unfertilized soils, in spite of losses from insects, and so forth. Naturally the yields of net food have to be less, and no amount of genetic breeding will help.

It is a cruel hoax to send so-called improved varieties which are in reality incomplete parts of the industrialized agriculture to underdeveloped systems where the part actually needed must have more useful work circuits within it. The man from the poor country, looking at affluence and being told it is due to improved varieties throws out a workable if low-yield system in exchange for an incomplete part of an agricultural system that cannot even operate without the full support of the industrial fossil-fuel inputs and management.

OCCUPATIONS AND SPECIES

Earlier we showed how natural systems are characterized by a network of specialists, each of which is a species. Figuratively, the species populating Noah's ark were necessary to provide a network capable of supporting man. Modern agricultural systems in industrialized societies also operate with populations of occupational specialists but here people replace many other species in the performance of work requiring a division of labor. In either case specialization by division of labor provides the overall system with a greater amount of power converted to the work necessary for continuance and competitive group survival. It is interesting that the number of occupations has been found similar to the number of species in natural systems. A comparison is given in Figure 4–11. Although gifted with the advantage of flexibility, the occupational groups in human societies develop some of the same insulation and functional separation that the species have in locations of their work, training, clothing, and communication.

As man receives auxiliary energies from fossil fuels, he uses less and less of the natural network of species but replaces them with work functions in his own society. Parsons [28] in recognizing society as a system identified such things as pathways, channels, circuits and insulation, and subsystems, although the terminology was different.

A SCALE OF NET YIELD AND POWER SUBSIDY

The various kinds of systems, from the primitive ecological one with man as a minor component to the more modern one with man securing food from petroleum subsidy of intensified agriculture, form a series shown in Figure 4–12(*a*). As we read the scale from left to right, the ratio of fossil-fuel supplement (or other special energy subsidy) to solar energy increases. The net yield that is possible after the work required for maintenance and survival has been carried out also increases. A similar graph was provided by Giles [12] for the fossil-fuel contributions in the fields (Fig. 4–12(*b*)). The relative positions of solar agriculture and some varying levels of intensified modern agriculture are indicated. If the accuracy of this diagram can be exactly established by more research, it should be possible to select an amount of fossil-fuel and the corresponding agricultural systems that can be used for a country with that measure of power support. If we retain the know-how for agricultural-ecological-human systems for each degree of energy supplement, we can recommend systems for the underdeveloped parts of the world. If it becomes necessary, we can change our own role too.

The number of species circuits needed to maintain necessary diversity is part of this scale, for we use these natural networks for survival on the left and replace them with circuits of man's specialty on the right. By this theory the number of occupations of people and the number of

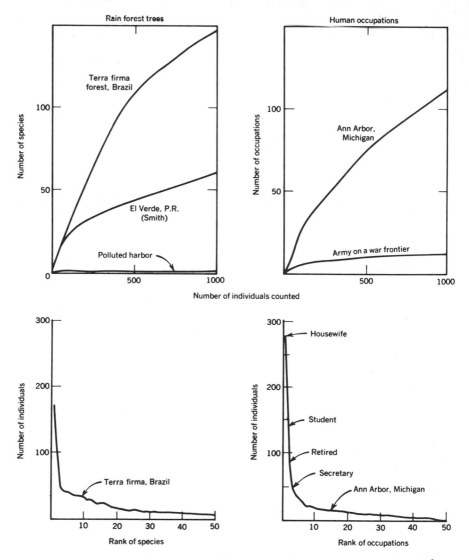

Figure 4–11 Patterns of occupational diversity in human systems at Ann Arbor, Michigan [7] compared with species diversity in natural systems without man, rain forest of Brazil [5] and mountain rain forest in Puerto Rico [31].

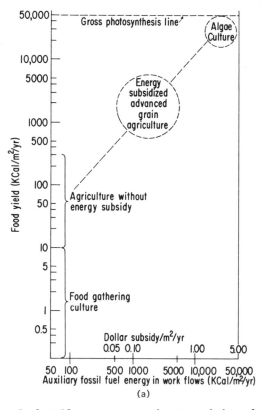

(a)

Figure 4–12 Net food yields to man as a function of the subsidy of fossil-fuel industry. (a) Diagram of the net yields possible for man from a photosynthetic system with auxiliary fossil-fuel energies being used to do work that would otherwise have to be done from the yield [25]. As favorable environmental supplements and fossil fuels become large, the net yield to man approaches the maximum gross photosynthesis ceiling inherently limited by the thermodynamics of photosynthetic cells. The ceilings for the gross photosynthesis of the best agriculture and the most productive natural communities are about the same, here written as around 150 kcal/(m^2)(day). (b) Relationship between yields in kg/hectare and power applied on the farm in horsepower per hectare [12]. Similar data are given relating fertilizer, another input from industry [23].

species in the system are complementary; cutting down one requires increase of the other.

FUEL CLASSIFICATIONS OF HUMAN SYSTEMS

The fuel-rich civilizations achieve much denser populations because they are running on the fossil fuels; as we said earlier, even the potatoes man

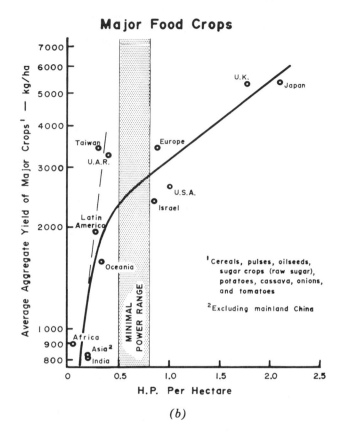

Major Food Crops

(b)

eats are partly made of oil. The larger carrying capacity is based on the continued flow of the concentrated potential energy. The uneven distribution of wealth is really an uneven distribution in the application of fossil fuels. Primitive areas with rich oil deposits can only sell the fuel because they have no network of advanced culture capable of accepting the subsidy directly.

In considering the problems of world food, Figure 4–12 may allow us to scale countries ranging from the most power-rich to those still surviving on solar energy alone. The top horizontal line, based on the results of a study of photosynthetic rates in complex independent natural systems, (such as rainforest) shows the gross fixation of solar energy in photosynthesis (before subtracting that used by nature in self-sustaining work) as a ceiling that agricultural production approaches but does not exceed. Without fossil-fuel supplement agriculture never exceeds the gross photosynthesis of the natural system. Even with fossil-fuel supplement, it does not exceed the gross photosynthesis rate in those optimal areas of the world where there are no sharp limiting factors other than light energy. Man may improve gross production only when there are limits such as water or mineral shortages that can be overcome with fossil fuels. In Figure 4–12 the top line corresponding to maximum gross photosynthesis, the unlimited natural system, and the best agriculture is drawn as around 150 kilocalories of organic matter fixed per square per day in the growing season.

The carrying capacity for man of a system based entirely on solar energy is of the magnitude of 0.00025 to 0.000025 persons per square meter with 99.5 percent or more of the organic matter production metabolized in the work of life support which includes the converging preparation of the 0.5 percent or less that enters the human as food. These figures are based on studies of existing societies of primitive man embedded in nature and of man working as a simple farmer at about one person per acre or less. Higher densities have occurred in special places where nature transported the production of larger areas to the region by means of a stream, for example, salmon passing in waterways. If we consider the real area of primary photosynthetic support, however, no greater human densities than those have been shown to be stable.

At the other extreme of the diagram, we see all the work drains of the support system being provided from outside energies so that the solar energies can go into human food. Thus, in heavily subsidized agriculture, the yields of food and fuels begin to approach the gross production ceiling. By figuring out the ratio of solar-energy budget to fossil-fuel budget, we can locate a country along this scale and thus know what system of food production it may be expected to have in excess of its needs.

As Brown [4] shows, the agricultural yields per capita and per unit of land have taken a sharp upward turn in the industrialized countries in this century, whereas the others have shown little gain in production and a loss in relative dollar value of their produce. He called this "take-off." The difference represents the fossil-fuel subsidies, as shown in Figures 4–6(a) and 4–7, which permit 97 percent of farm production in the United States to be exported to the economy. Notice the bends in the lines in Figure 4–12. Food production is increased but the amount added per dollar of industrial subsidy has decreased.

DIRECT CONVERSION OF FOSSIL FUEL TO FOOD

We may conclude this discussion of the energetic basis of man by considering some currently active proposals for making food directly from petroleum without any solar energy. Man cannot eat fossil fuels because the fuel molecules are not digestible, but many microorganisms readily digest the hydrocarbons. New microbial systems are being tested by the oil companies (see Figure 4–13). Based on an estimate by McPherson [22] and yield data provided for a methane system by the Shell Development Company, Figure 4–13 shows conversion of fossil-fuels into bacterial tissue and hence food. One disadvantage may be that direct use of fossil fuel alone is less efficient than the sum of fossil fuel and some solar power. The efficiency in the example is around 10 percent as in natural food chains. Using fossil fuels in this way rather than in the control and elimination of natural limiting factors produces less food per fuel calorie, as we can see by comparing Figures 4–7 and 4–13. Much higher efficiencies are claimed in slow batch experiments by Johnson [17].

Even more basic is the direct conversion of oil to food molecules such as sugar by direct organic chemico-industrial manufacture. Some precedent for synthetic food manufacture may be cited from World War II when Germany, under the stress of failing food supplies, oxidized petroleum to make an edible fat.

NUCLEAR POWER BASE FOR MAN

At present the expansion of nuclear power suggests no immediate great boost to the energy economy although nuclear power may substitute for fossil fuels at similar levels of net power yield. Rising requirements for safety and for waste handling tend to cut into the net yields of nuclear power, already heavily subsidized by the fossil-fuel economy through

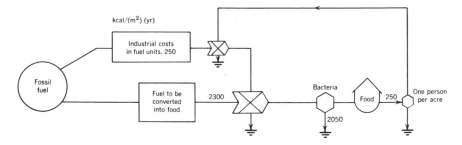

Figure 4–13 System converting methane and/or petroleum to food using industrial-microbiological means of growing bacteria on hydrocarbons. Calculations are prorated to indicate fuel requirements per area to support 640 persons per square mile. The fuel cost of the industrial subsidy is computed from the $0.55/kg cost estimated by McPherson [22]. The percent of fuel converted was obtained from a report on a Shell Development Company pilot plant in which 10^7 grams of food (50 percent protein) was produced from 2 million ft³ methane (4.6×10^8 kcal) [26].

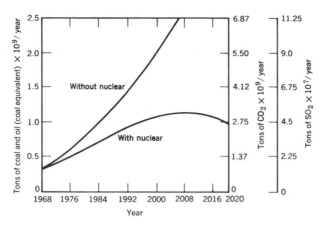

Figure 4–14 Past and future electrical power development using fossil fuels and nuclear power from Tape [32]. Also shown is the metabolic carbon dioxide and acid sulfur releases equivalent to power generated.

the Atomic Energy Commission and industries. Whereas official extrapolations of nuclear power expansion by the AEC suggest that the fuel is a net gain comparable to opening new fields of easily accessible oil [32] (see Figure 4–14), the statements of skeptics such as Ehrlich and Holdren [10] raise doubts. The discovery of Arctic oil is certainly as important at present. The nuclear power certainly allows a safety back-up and perhaps a stable power period as accessible oils diminish.

No one really knows the net yield of nuclear power because at present its use is subsidized by fossil fuels in a thousand ways that cannot be estimated until we try to run a nuclear system without them.

Will nuclear power have a more concentrated value than the wood output of the solar system, or of coal, or of cheap oil from rich deposits? The new power plant seems to be more economical than the competing fossil plants as long as it is running on the accumulated storages of nuclear fuel and fuel prospecting done on fossil-fuel subsidy. Is nuclear power at this level of net power delivery possible in a culture that does not have the accompanying fossil fuels?

At present the nuclear power goes with highly developed fossil-fuel economies rather than as the means for building a separate economy in an underdeveloped country. We cannot speak at present of a nuclear power culture. We might imagine its nature if it develops as even more centralized large industrial complexes that concentrate around the large nuclear power units and a disappearance of the many decentralized flows of excess power such as the automobile. In this respect a more sensible pattern of life may develop. Also the metabolic wastes may be less.

If however, the trend towards miniaturization follows, the new technology may link radioactivity to electricity in small units such as the ferro-electric battery. An increased general level of radiation is really an increased level of error and may be the same in the long run as other kinds of pollution, both requiring increased energies of medical maintenance. A more resistant population may develop by natural selection.

SUMMARY

With different sources of power and different roles for man, the systems spend different quantities of potential energy on his life-support system and hence on him. In small numbers in deep forest the area per man is large and much photosynthetic energy forms the base of his life support. Modern man with large fossil-fuel expenditures as well as solar energy forming his life support requires much more than a man in monsoon agriculture based on rice in Southeast Asia. These figures are 100 to 1000

Table 4–2   Power Requirements Per Day for Total Life Support of a Man

	Area Per Man (Acres)	Solar Energy (kcal/day)	Organic Matter [million kcal/(day) (person)]
Pygmies in deep forest, solar base[a]	640	10^{10}	341
Man, if evenly distributed over the earth	25	10^9	0.4
Man in United States	12	10^8	
Fossil-fuel base[b]			0.6
Photosynthetic input[c]			1.0
Total			1.6
Man in solar-based, monsoon agriculture[d]	1	10^7	0.16
Man in United States city[e]	0.0064		0.1
Man in Apollo space capsule	0.001		
Life support[f]			124,000
Capsule[g]			2,740,000

[a] 1/square mile; forest production 131 kcal/(m^2)(day); 2.6 × 10^6 m^2/square mile.
[b] 1.2 × 10^{14} kcal/day; 200 million people.
[c] 54/acre; 40 kcal/(m^2)(day) for half of year.
[d] 1/acre; 4 × 10^3 m^2/acre; 80 kcal/(m^2)(day) for half of year.
[e] 10^6 people; 100,000 people/square mile; food 2000 tons per day; fuel, 9,500 tons per day [35]. Not included are the fuels used outside the city in its support.
[f] Fuel consumption rate after in space, 9.4 kilowatt [16].
[g] $$10^6$ capsule cost prorated over 10 years; 10^4 kcal/$.

times greater than the 3000 kilocalories of high-quality nutrition per man that is required to operate his body daily. Man's life support requires much more than his food; he requires a system in which he is compatible whether it is manufactured and maintained by the forest, by sunlit fields, or by fuel-eating machines. Table 4–2 provides a summary. Monsoon man has a relatively small energy base. Forest man has a large energy base, providing a stable and a large area of beauty and interest, but little of it reaches him in any other direct way. United States man spends large quantities of high-grade potential energy in his support system, also converging the output of many acres of solar energy to support each man. The costs for man in space are astronomical.

REFERENCES

[1] Anonymous, *World Food Budget*, U.S. Dept. of Agriculture, Foreign Agricultural Economic Report No. 19, 1970.

[2] Alkire, W. H., "Lamotrek Atoll and Inter-island Socioeconomic Ties," *Illinois Studies in Anthropology*, No. 5, University of Illinois Press, Urbana, Illinois, 1965.

[3] Birdsell, Joseph B., "Some Environmental and Cultural Factors Influencing the Structuring of Australian Aboriginal Populations," in J. B. Bresler (ed.), *Human Ecology*, Addison-Wesley Publ. Co., Inc., Reading, Massachusetts, 1966, pp. 51–90.

[4] Brown, L. R., *Increasing World Food Output*, Foreign Agr. Eco. Report No. 25, U.S. Department of Agriculture, Economic Research Service, Foreign Regional Analysis Division, 1965.

[5] Cain, S. A., and G. M. De Oliveir Castro, *Manual of Vegetation Analysis*, Harper, New York, 1959.

[6] Carter, W. E., *New Lands and Old Traditions*, Latin American Monograph, No. 6, Univ. of Florida Press, Gainesville, 1969.

[7] Clark, P. J., P. T. Eckstrom, and L. C. Linden, "On the Number of Individuals Per Occupation in a Human Society," *Ecology*, 45, 367–372 (1964).

[8] Davis, D. H. S. (ed.), "Ecological Studies in Southern Africa," *Monographie Biologicae*, Vol. 145 (1966) (Dr. W. Junk, Publisher, The Hague).

[9] Deshler, W. W. "Native Cattle Keeping in Eastern Africa," in Anthony Leeds and Andrew P. Vayda (eds.), *Man, Culture, and Animals*, Publication No. 78, American Association for the Advancement of Science, Washington, D.C., 1965.

[10] Ehrlich, P. R., and J. P. Holdren, "Population and Panaceas, a Technological Perspective," *Bioscience*, 19, 1065–1071 (1969).

[11] Fischer, A. W., "Economic Aspects of Algae as a Potential Fuel," in F. Daniels and J. S. Duffie (eds.), *Solar Energy Research*, University of Wisconsin Press, 1961, pp. 185–189.

[12] Giles, G. W., "Agricultural Power and Equipment," *The World Food Problem*, Vol. III, A Report of the President's Science Advisory Committee, White House, Washington, D.C., 1967, pp. 175–208.

[13] Hagen, E. E., "Man and the Tropical Environment," Symposium on Biota of the Amazon, Mimeo, Belem, 1966.

[14] Harris, M., "The Myth of the Sacred Cow," in Anthony Leeds and Andrew P.

Vayda (eds.), *Man, Culture, and Animals,* Publication No. 78 of the American Association for the Advancement of Science, Washington, D.C., 1965, pp. 217–228.

[15] Hickling, C. F., *Tropical Inland Fisheries,* John Wiley and Sons, New York, 1961.

[16] Jenkins, D. W., "Biogenerative Life Support Systems," *Bioregenerative Systems,* National Aeronautics and Space Administration, NASA, SP-165, 1–6 (1968).

[17] Johnson, M. J., "Growth of Microbial Cells on Hydrocarbons," *Science,* **155,** 1515–1519 (1967).

[18] Kleiber, M., *The Fire of Life,* John Wiley and Sons, New York, 1961.

[19] Lawson, G. W., *Plant Life in West Africa,* Oxford University Press, London, 1966.

[20] Leeds, A., and A. P. Vayda, (eds.), *Man, Culture, and Animals,* Publication No. 78 of the American Association for the Advancement of Science, Washington, D.C., 1965.

[21] May, J. M., *The Ecology of Malnutrition in Middle Africa,* Hafner Publishing Co., New York, 1965.

[22] McPherson, A. T., *Food for Tomorrow's Billions,* Proceedings of "Food in the Future Concepts for Planning," sponsored by Dairy and Food Industries Supply Association, 1145 19th St., Washington, D.C., 1964.

[23] Nelson, L. B., and R. Ewell, "Fertilizer Requirements for Increased Food Needs," *The World Food Problem,* Vol III, A Report of the President's Science Advisory Committee, White House, Washington, D.C., 1967, pp. 95–118.

[24] Odum, E. P., *Fundamentals of Ecology,* 2nd ed., Saunders, Philadelphia, 1957.

[25] Odum, H. T., "Energetics of World Food Production," *The World Food Problem,* Vol III, a Report of the President's Science Advisory Committee, White House, Washington, D.C., 1967, pp. 55–94.

[26] Overbeek, J. van, personal communication.

[27] Parker, F. W., *Food for Peace,* American Society of Agronomy, Special Publication No. 1, 1963, pp. 6–20.

[28] Parsons, T., *The Social System,* Free Press, Glencoe, Ill., 1951.

[29] Petrides, G. A., and W. G. Swank, "Population Densities and the Range-Carrying Capacity for Large Mammals in Queen Elizabeth National Park, Uganda," *Zoologica Africana,* 1, 209–225 (1964).

[30] President's Science Advisory Committee on World Food Production. *The World Food Problem,* White House, Washington, D.C., 1967.

[31] Smith, R. F., "The Vegetation Structure of a Puerto Rican Rain Forest Before and After Short Term Gamma Irradiation," *A Tropical Rain Forest,* U.S. Atomic Energy Commission, Division of Technical Information, Oak Ridge, Tenn, in press, pp. D-103–D-135.

[32] Tape, G. F., *Environmental Aspects of Operation of Central Power Plants,* U.S. Atomic Energy Commission, Division of Technical Information, pamphlets SP -54–68. 1969.

[33] Turnbull, C., *The Forest People,* Simon and Schuster, New York, 1916.

[34] Turnbull, C., "The Lesson of the Pygmies," *Scientific American,* **208,** 1, No. 1, 28–37 (1963).

[35] Wolman, A., "The Metabolism of Cities," *Scientific American,* **213,** 179–190 (1965).

[36] Drummond, A. J. "Radiation and the Thermal Balance," UNESCO, *Climatology,* 56–74 (1958).

5

POWER FOR ORDER AND
EVOLUTION

THE PERPETUAL SHAKEDOWN

By now most educated citizens know that the random jumping of warm molecules in all matter, plus the many kinds of frictions in the world of larger systems, invariably tends to shake order into disorder. The higher the temperature, the faster the little structures shake apart; and the more weathering processes, mechanical wind processes, destructive radiations, and earth movements there are, the more the large structural patterns become disrupted. The loss of order into disorder is another manifestation of the principle of energy degradation.[1] When some of the small molecular structures shake apart as the molecules wander at random, the effects may be very large energetically because there are so many molecules in a small volume. When some of the patterns in the big structures, such as arrangements of parts in a stock room or the patterns of a fossil bed, shake apart, the energy involved may be relatively small because

---

[1] This principle is also stated as: the universe tends to run down; molecular patterns trend from the less probable to the more probable. In dispersion of energy as discussed in Chapter 1, the wandering of molecules from some less probable situation to one more probable is a heat dispersion. The ratio (Q/T) of heat dispersed (Q) to the temperature (T) is called change of entropy. The molecules wander so as to even up the intensity of their motion (temperature), and the ratio in the section from which the molecules wander goes down more than the ratio goes up in the section into which the molecules wander. Consequently, the overall ratio always increases, counting the whole process of wandering. Thus this entropy ratio is often said to measure the amount of degradation going on.

the pattern of arrangement, although large in dimension, is thinly spread over the matter arranged, at least in comparison with the concentration of order found in such associations of chemical molecules as coal, oil, or dynamite.

MAINTENANCE OF ORDER

With the continual shakedown of order, we cannot keep structures organized unless we continually restore order. Just as we continue to repair an automobile, we must continually repair all desired structures in the small and large patterns of the earth. All living structures can hold their form and function because they have the capacity for self-repair. Maintenance involves several kinds of processes, all of which must use some of the flow of potential energy into dispersed heat to accomplish their purposes. Some energy goes to reorganizing those units whose arrangement has come awry. Other energies go into restoring energies that have been depleted from necessary points of storage and force regulation. Organisms have a continuous cold fire (biochemical reactions with the energies coupled to do work) that shows that their maintenance is continuous. When an animal or plant tissue is doing nothing except resisting the shakedown, its rate of burning fuel is called the basal metabolism. The way in which some of the power flow is turned into repair is indicated for organisms in Figure 3–13 and in general terms in Figure 5–1(a).

A self-maintaining population arranges for inflow of materials (2) and potential energy (1); combines materials to form parts (3); throws old parts out (6), sometimes reusing them as materials; rearranges new parts and disarranged situations (4); and transforms fuel energies into new storages of potential energy when the form must be different to operate the system (5). If the new replaced parts are separately visible entities, we call their formation reproduction. There is net gain when the balance of input and outflows of energy increases storage at 7.

Whether we are discussing populations of one species, of cells, of communities, or of other self-reproducing groups, similar processes are involved and may be collectively indicated with the hexagonal symbol shown in Figures 3–13 and 5–1(b). Coon [3] describes humans as connectors who transform potential energy into social structure.

HOW MUCH REPAIRING?

Much is known about the metabolic repair systems of living organisms. In general, the more organisms there are to maintain, the more power

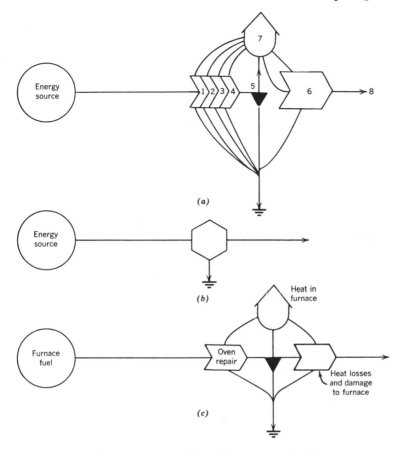

Figure 5–1 Energy flows, storages, and work process involved in maintenance. (*a*) Components of maintenance: (1) fuel processing; (2) material processing; (3) synthesis of parts by combining materials; (4) rearrangement and connections of disarranged parts; (5) energy storing for necessary fuel reserves and necessary structure; (6) removal of old worn parts; (7) energy storage; (8) exported products. (*b*) Summarizing symbol for a self-maintaining population with the functions 1–7. (*c*) A furnace with stored heat looped back to useful work of repair.

flow must be diverted to do the repair. Thus if we start to increase the population of organisms or the amount of desired and necessary structures, we must have more power flow.

Aspects of repair, replacement, inherited information, and evolution are diagrammed in Figure 5–2. Among higher organisms the pattern of survival includes some temporary repairs (Fig. 5–2(*a*)), but generally a deterioration of structure is allowed followed by a discard at death,

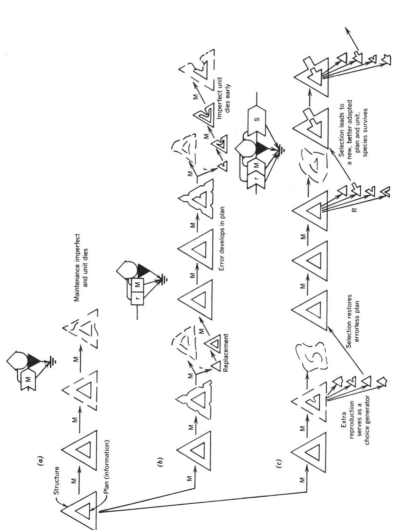

Figure 5-2 Diagrammatic representation of transmission of structure with time. (*a*) Maintenance work: structure is maintained, but imperfectly, so that eventually the unit dies. (*b*) Maintenance and replacement: structure is maintained and one unit is reproduced to replace the one that wears out, but eventually errors develop in the plan used for reproduction and the unit becomes imperfect and dies. (*c*) Maintenance, replacement, and loop selection of extra choices: structure is maintained and also has reproductive replacement. There is the additional work of maintaining an error-free plan or generating an improved plan by combining extra reproduction-generating choice and selection that benefits them.

142

the new individuals being made anew with mass production processes (Fig. 5-2(b)). Apparently making new organisms is energetically cheaper than repairing the old ones. Repair is not a mass production procedure. The pattern is similar in this respect for people, oaks, and automobiles. In considerable contrast, the associations of separate organisms do not show such senescence, and renewal does not require total replacement. The parts of the systems are like organisms and are individually replaced as they age, but the system of parts is ordinarily repaired rather than replaced.

One possible explanation for the difference in means of maintenance is the difference in geometry. For organisms and cars the parts are attached and intricately connected so that repair by rearrangement, or by removal and replacing a part, is difficult. In the open ecological system the organismal parts are not intimately connected geometrically; therefore, they may be separated and repaired one at a time relatively easily. Thus senescence and replacement of the whole may be economical when the parts are intimately connected, whereas repair by parts replacement and potential immortality may be the pattern when the parts are not intimately connected.

Whereas most loose associations of mobile organisms are not intricately connected, monolithic rigid structures do develop where power flows are highly concentrated. Examples of unconnected associations are the organisms in plankton of the sea, the scattered plants in a new field, or the houses in suburbs. Examples of connected associations are the reefs of oysters, heavy root networks of some tropical forests, or the continuous apartment complexes of cities. It is the latter group that develops senescence at the group level as well as in its parts. We are used to the idea in urban renewal that some continuous building structures are more cheaply replaced than repaired. An example in a simpler ecosystem is the senescence of barnacle associations illustrated by Barnes and Powell [1] in Figure 5-3. When old and top-heavy they break off or are broken off by animals that serve an urban-renewal role in the animal city. New growth and succession refills the gaps.

Senescence apparently only occurs in those physically attached units of such complexity that the cost of disengaging parts for replacement becomes too high. Senescence is a property of the compact and complex.

LITTLE SHAKING IN A BIG CORRAL

The generality just stated, that power is needed in proportion to the amount of structure to be maintained is true as long as we are discussing

(a)

(b)

Figure 5–3 Associations which have monolithic structure and develop senescence. (*a*) Cross section through crowded barnacle growths from Barnes and Powell [1]; (*b*) urban ghetto where houses have common walls.

structures of the same general size. If, however, we wish to maintain order of a tiny structure, we must repair it more often, for the molecular shaking² can break it apart readily, the structure being only slightly larger in size than the molecules doing the shaking. In fact, we cannot marshal and focus enough energy to make and maintain a structure the size of these shaking molecules.

² Most readers will remember seeing demonstrations of the Brownian movement through the microscope in their science courses; in such demonstrations visible particles smaller than bacteria can be seen to shake because of the bombarding and wiggling of the invisible molecules around them.

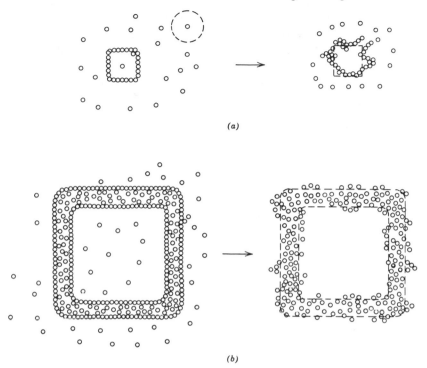

(a)

(b)

Figure 5–4 Comparison of the relative loss of form of large and small squares through random molecular motion. (*a*) A small square in which the action of one molecule constitutes a large proportion of the form; (*b*) a large square in which the random motions of single molecules are less important and the form lasts longer without repair.

A structural form can be built and maintained more inexpensively by fashioning it of large blocks (Fig. 5–4). The molecules may wander and disorganize within the small blocks without ruining the desired structure, which is on a larger scale or is in the relationship of the big blocks. Thus we see such different structures as buildings of bricks or bodies of elephants with lower maintenance costs per pound than those of microscopic structures such as electrode metal surfaces and microorganismal bodies. The size-maintenance story is the reason the increase of randomness in energy flows at microscopic levels can be used to maintain structures of greater dimension.

An analogous system is a fiery furnace in which some of the energy goes to help repair the furnace. (Some steel-making furnaces rely on continual heat effects to condition the walls.) The energies involved in the molecular randomizing in the furnace fuel are so vast compared to the maintenance costs of the walls of the big furnace pattern that one can support the other while throwing its energy down the heat-dispersion drain as required by energetic law. The energy diagram for the furnace is similar in these respects to that for a living system (Fig. 5–1).

SCHRÖDINGER RATIO AND METABOLISM

At one time in this century the concepts of increasing randomness in physical science appeared to indicate a trend against the facts of order and progress in living systems. This misconception developed partly because some of the physical scientists were not taking much biology and knew very little about the old facts and theories of basal metabolism and maintenance, and biologists were similarly little exposed to physics.

An across-field synthesis was made by Schrödinger [11], putting basal metabolism into the language of energetics (Fig. 5–5). Refrigerator boxes are made colder by a process of pumping much more heat from the power source into the environment than is removed from the box as required by the energetic degradation principle. Schrödinger used the same point that the high degree of order of biological tissue is maintained by pumping much more potential energy of fuel power into the environment than the heat[3] randomness removes from the structure. Schrödinger used the

---

[3] Heat is the molecular shaking motion; the level of molecular shaking is called temperature. When a person feels hot, his molecules have energy, are shaking more, and indicate this to the person's conscious self by a sensory system. As the shaking decreases, we say the temperature goes down. At a certain temperature, which is almost reached in some laboratory work, the shaking is zero and the thermometer reads $-273°$ C; temperature cannot go below this point, called *absolute zero*. The molecules are still and are said to be ordered. The *absolute temperature scale* starts at absolute zero and thus is at $+273°$ C at freezing of water.

If we start adding some of the shaking energy (heat) to material initially at absolute zero, the temperature rises. If we sum up little by little the heat calories added and divide each amount by the temperature at which it was added, according to the entropy quotient formula described earlier, we obtain a gradually increasing quantity which measures how much shaking would have to be removed again to restore the complete order of absolute zero.

Thus, the final quantity determined by such calculations is the *entropy content* of the structure. To lower the temperature in one place, we must raise the surrounding temperature even more in order to follow the energetic principle of degradation. Thus refrigerators become cold by pumping vastly larger quantities of heat out into the room. The entropy in the box goes down, but the overall entropy total goes up as mentioned earlier.

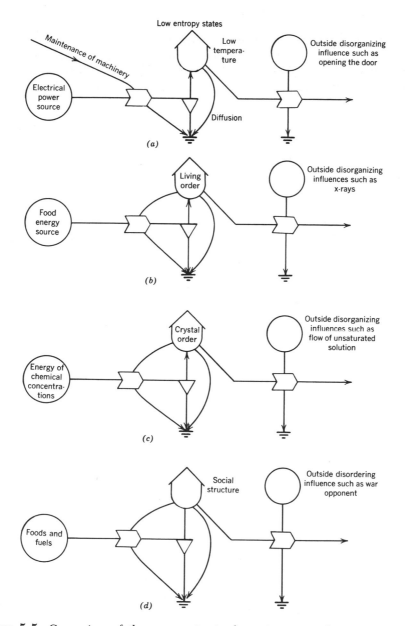

Figure 5-5 Comparison of the energy circuits for maintaining a low temperature in (a) a refrigerator, (b) the ordered structure of a biological system, (c) a crystal, and (d) a social system. All are in states of low entropy. Potential energy is dispersed into heat-generating entropy in the Schrödinger maintenance process and in the disordering outflows.

147

entropy ratio as his language for stating this principle. Compare systems (a), (b), (c), and (d) in Figure 5-5. Refrigerators, crystals, tissues, and societies have a relatively ordered structure (improbably low entropy content) which is maintained by continually pumping out into the environment large quantities of dispersing heat so that the overall entropy ratio always goes up. In the nonliving system we can obtain order by means of a low temperature (less shaking) and its maintenance; in the living system order is maintained at normal temperature by counteracting with maintenance work the shaking actions that disrupt them.

Part of the maintenance protects the surfaces of the energy storages from reactions with the outside in the same way that walls protect grain storages and paint protects cars. These requirements are less per pound in large bodies because the surface to volume ratio is smaller. Metabolism for this can be and is less. Elephants have less skin to maintain per pound than the bacteria. If storing is needed, big tanks have economical costs.

The potential energy stored measures the reaction tendencies to be averted and thus the ordered structure being protected. Usually the potential energy we measure in a mass is relative to reaction with outside oxygen. Even when we give calorie values for a living substance we imply its potential energy as a homogeneous mass of organic fuel molecules in reactions with oxygen. Shaking and shuffling of the molecules in a homogeneous fuel tank does not affect its order or energy value.

More essential to life and measured with more difficulty, however, are the potential energies between the tiny microscopic structures of the living protoplasmic parts relative to each other and to substances flowing through the living processes. Maintaining the multitude of these tiny structures and with them the potential energy differences on a microscopic scale is a primary job of respiratory work of living protoplasm (basal metabolism). The calorie values of the protoplasm in bulk relative to outside air do not measure these important but smaller internal gradients of energies. When a living organism dies, disordering reactions no longer negated by the living metabolic work, shake the living structure apart quickly. Although not measured directly we use the respiration as a measure of these structures to be maintained.

The Schrödinger statements imply that there is some law about the amount of power flow that has to go into the drain (increasing external disorder and entropy) to keep each unit of structure ordered against its thermal shaking and other disorganizing effects. The metabolisms of organisms provide the power demanded for maintaining ordered states. The total power flows in nonchanging ecological communities or in stabilized cities of man can be similarly used to learn the power flows necessary per unit structure of different sizes, forms, and energy storage.

DARWINISM AND THE CORRECTION OF ERROR

To repair and make new parts, we must use templates and plans just as is done in the auto maintenance industry. The loopback of power flow into maintenance is accomplished as required by energetic law, but this energy supply does not explain how something can be repaired when even the templates and plans are shaking apart too. The systems that do repair work in many organisms are not intelligent beings. As we shall see, even the plans for intelligence, networks, and other large patterns are maintained through a wonderful correction system recognized by Darwin. In Figure 5–2 loop selection of extra choices of reproduction is diagramed. The concept of natural selection involves the choice by nature from among alternatives provided. The making of more offspring than can survive provides the opportunity for those most suited to surrounding conditions to last and in turn produce more offspring. If all the offspring survive, there is no selection; a choice must be provided by excess creation and a system of selection.

In energetic terms, the putting out of excess offspring is a power-demanding process that can be measured in terms of the energy flow required for the total metabolisms of the reproductive process and of the offspring during their existence. The energy cost of the selection is the power flow involved in the mechanism that arranges for the continuation. The mechanism for selection in simple organisms may be the greater efficiency in the loop recycling materials, including continual absorption and remineralization of the excess offspring into the biological cycles.

In higher organisms, such as carnivores, behavior systems for specializing selection are set up. In man the introduction of will, consciousness, and group-favoring motivations (religious, economics, egoistic, etc.) further specializes the selection of subsystems. These mechanisms may increase the power costs.

Thus on the one hand we have the shaking and other disordering forces and on the other hand the process of proffering more duplicates than are needed and selecting the best to serve as the templates for the next round. As fast as errors develop in the templates, the selective process takes them out. If too little power is put into superduplicating and selection, the errors will exceed the repair and the stock will lose order. Systems with loss of order may still survive at a lower level of structure if balance of disordering and ordering is again achieved.

LOOP REWARDS AND NATURAL CURRENCY

Not only must the selection continue those populations most suited to the system's efficient function, but it must also pick those that properly contribute to the total long-range order and energy flows of the system in which it is embedded. Cancer and the explosive growth of some exotic transplanted weeds are examples of inharmonious intruders, growing too fast in relation to the flows of material and energy capable of long-range support. Man's present civilization may also resemble a weed growth. The system impresses its control on the part by *loop circuits* which recycle materials that reward and stimulate in proportion to the effectiveness of the parts for the system. Either too much or too little growth by one component puts the loop out of harmonious flow.

Although the details are not yet certain in the world of giant molecules in cells, duplications are made from templates. For example, the nucleic acid chemical codes on chromosomes are transmitted to messenger bodies which direct the manufacture of the enzymes; these in turn serve as valves on chemical reactions. If there are routes for positive stimulation from the most functional enzymes back to the template duplication, these routes constitute system control and Darwinian selection in a primitive form. Just as money in society serves as a positive feedback stimulus of those outputs favorable to its needs, so the cycling of materials serves as a system of loop currency, rewarding pathways that become circular with further use.

LOOP SELECTION PRINCIPLE

In ecological studies there is the positive feedback loop through which a downstream recipient of potential energy rewards its source by passing necessary materials back to it. For example, the animals in a balanced system feed back to the plants in reward loops the phosphates, nitrates, and other compounds required for their growth. A plant that has a food chain which regenerates nutrients in the form it needs is therefore reinforced, and both plant and animal continue to survive. Species whose work efforts are not reinforced are shortly eliminated, for they run out of either raw materials or energy. They must be connected to input and output flows to survive.

In man's complex system he has arranged a feedback currency that is even more fluid than the geochemical recycling of natural ecosystems. He invented money, which is fed back in reward for work done. The flow of each population is thus looped to at least one other population, and by interconnecting loops the economic system provides rewards for each and a means by which the system designs the parts. In the same manner that learning occurs in the development of children, reinforced and reused loops survive, replacing the possible alternatives that do not become reinforced. Nature's systems design themselves by this evolutionary mechanism. The networks of man use similar means. Understanding the role of closed-loop selection allows us to apply the principle in our development of systems which can survive.

POWER COST OF EVOLUTION

In evolution, not only are the structures and patterns of the systems maintained against the disordering influences through duplication, selection, and loopback currency, but, by definition of the term, novel changes are made. The power cost of the change is the power cost of making even more choices and selections than are required to hold the order stable. Thus a system can evolve new patterns more rapidly when there are more power flows, providing excess quantities of choice in the form of extra offspring, more specialized mechanisms for selecting, and more reward cycling. Without power excess change in the direction of new order is at times impossible, although loss of structure is possible.

The energetic facts of evolutionary change, whether of primitive organisms or of man's complex society, may explain much about the rates of evolutionary change, including the explosive transformations brought about by the great fossil-fuel injections in the past hundred years. During much of the millions of years of evolution, energy sources were evidently dilute, consisting principally of the sun, so that evolutionary innovation had to be slow. The life we now have represents the stepwise saving of very tiny increments of energy transferred into small increments of order. What percent of the power budget of an organism can go into innovation without inviting displacement by another system that puts all its calories into current needs and applies little to the future? If the environment is a fluctuating and changing one, the power budget may be shifted to increase the duplication-selection, but in times of stability such budgeting might be antiselective.

Many have dramatized the great complexity of the first life system and the amount of order required to operate any living units now known. The energetic facts state plainly that a long time was required to achieve

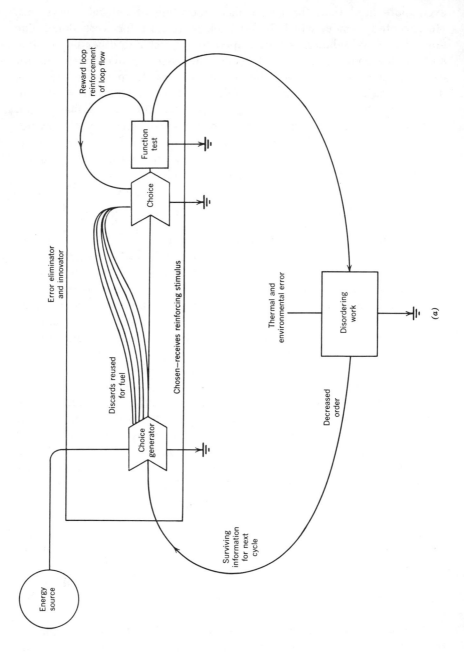

the present complexity of life through the step-by-step contributions from power duplication and selection over and beyond maintenance. Many scientists studying microscopic processes have not seen how selection could come about before there was life, possibly because they were not aware of the role of looping cycles in providing creativity in environmental systems in stages.

CREATIVE WORK MODULE FOR CONVERTING POTENTIAL ENERGY INTO MACROSCOPIC ORDER

We may express the principle of creativity and error correction in a general way, using the energy diagram in Figure 5-6. First there must

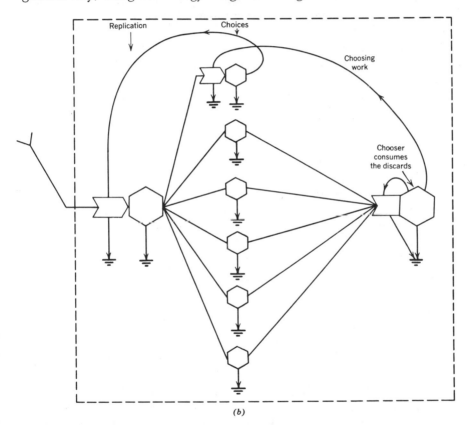

(b)

Figure 5-6 General diagrams for a choice-loop-selector for generating new arrangements through creative work. (a) The components include a random-choice generator, a tester loop for evaluating suitability to the system, and a selector unit. (b) A creative module in a prey-predator system.

be a variety generator and then a selector and, for economy, duplicates not chosen may be reused in energy and material processes. The energy expended as dispersed heat performs the work of generating the choices and supporting the chosen.

A nonliving example can be observed in cumulus cloud formation on sunny summer days. The heating of the ground develops at random many small rising currents which generate small competing cumuli that draw on the energy of the stored moisture. The ones that receive the most loop reinforcement draw more air flow and thus more energies, causing the demise of the others. The process of selection may develop a few large-scale thunderstorms by the end of the day. The rising columns are parts of loops of air that pass in down currents back to the ground constituting a mechanism that reinforces the cloud over the rising column and helps it move higher and increase in volume. The system disperses because its initial energy source is discharged. This is a natural selection process with a randomizing choice generator succeeded by a choosing system.

Another example under study by geologists [8] is the development of stream network order through the water flow driven by the potential energy of rain. The authors show how at first choice is generated by random development of many small tributary branches. Later, as these randomly developing drainage patterns join each other, some are reinforced and others pirated. The cutting of the ground allows the big ones to be reinforced and takes energy away from the smaller ones. The drainage pattern in the land represents order that is looped back to the future as memory unless the environmental disturbances upset the established tributaries.

In biological population maintenance or in evolution the choice generator is the excess reproduction and the choice machine may be the carnivores (see Fig. 5–6(b)). In organized reproductive systems the random mutations and reassortments of genetic mechanisms are the choice generator. In development and progress in industry, the choice generator is the research division and management serves as the selector.

In the embryological development of blood vessels or enzyme systems, the initial proliferation of many possible connections more or less at random constitutes the choice generation; the reward of those that manage to find a loop reinforcement of successful circular flow provides the selection that draws resources from the others and reinforces the chosen. Many evidences of adaptive growth within organisms are given by Goss [6].

In creative thinking the hours of mental tests and trials constitute the choice generator. Most of the ideas fail to receive the encouragement of striking a truth circuit reward, but a few are reinforced. Creative thinking therefore requires an effectively motivated random-thought generator plus a well-developed truth-reward response. Persons who are not creative may be impatient with the necessary continual unrewarded probing, may lack a random-thought generator, or may not have enough pleasure sensation when the ideas make a truth loop connection. Creative children with their half-organized random-thought generators often devastate the teacher's desired classroom order and receive the discouragement of negation; thus the development of their truth loop reward mechanism may be stunted.

Many kinds of creative special-purpose learning computers have been made each with a random varying device followed by a selection process which fixes those pathways which match some outside criteria [10].

In plant succession the multiple seeding and initial colonization constitute the choice generator, whereas the mineral cycles and developing food chains provide choice machines for rewarding the plants with loop development.

There is a possibility that such a creative process exists within single cells. If the template is continually replicated by some kind of loop from chromosome to protein template and the new chromosome is then reconstituted from the protein, these steps would make up a creative loop, including the opportunity for random generation and loop selection.

The diagrams in Figure 5–6 also serve to illustrate the normal process of science. Many kinds of trials called hypotheses are attempted, but the choosing system is the empirical measurement that makes certain the concept is loop-rewarded by some agreement with nature. In this way the network of theory grows, joining empirical facts. It is the empirical choosing system that sets science apart from philosophy. Ideas by themselves which do not receive loop reinforcement from real measurement wander off into interesting but unreal patterns. Both work units are required, the idea generator and the chooser.

REPLACEMENT ENERGY AS A MEASURE OF VALUE

In this book there are several uses of energy to estimate value, each having specific calorie numbers and each formalizing a different aspect of our common concepts of value. We have already considered the value of an energy source measured in its calories of potential energy stored or delivered per unit time relative to the general energy levels of the biosphere. The situation of quality food production illustrates another use of energy to measure value. Where potential energies work against friction and other processes and do necessary rearrangements, the product may be evaluated by the energy replacement cost even though the energies which did the work have passed out of the system. In the example of food processing, we are used to the high costs of middlemen, a necessary step in quality nutrition outlets (Figs. 3–9, 4–8). Since most natural systems do many operations at once, it is still difficult to isolate the part of the energy flow associated with the process of concentration. The costs of such processes as concentrating chemical diversity are less when the system of processing is shared with other useful operations.

The replacement energy cost increases as more organization and information is specified in the assemblage of increasing value. If this structural development increases the effectiveness of the object in amplifying other energy flows, its input value to other energy reactions also increases. In Chapter 6 we see these are related through the concept of balance-of-work payments.

For complex units with a budget for self-maintaining respiration balancing the steady deterioration rates inherent in structure, time becomes equivalent to energy, the energy of existence work during that period. Hence any time-delaying aspect of a system can be given an energy flux value as a measure of the loss (stress).

THE ECOLOGICAL SYSTEM PRECEDES THE ORIGIN OF LIFE

Many systems of energy flow drive circulating matter and hold patterns of structure orderly by pumping their potential energy budgets down the energetic drain. Principal examples are the world's wind systems, hurricanes, the water systems of rivers and oceans, and the cycle of erosion, deposition, and mountain building, and they all run on solar energy. As long as there is a steady power inflow of potential energy such as sunlight and as long as there are closed circuits of minerals, a system has the ingredients necessary for duplication, selection, and currency loopback of materials to the energy receivers. As shown in Figure 5–7, the circulation of seas provides a route for materials to pass from a photochemical zone by using energy and active molecules established in the light zone. The essence of the modern system of P and R existed before life. Figure 5–7 suggests some steps in the chemical evolution toward life.

The flows that favor a closed loop receive the currency of raw materials from the down-energy part of the system. Thus flows of the normal

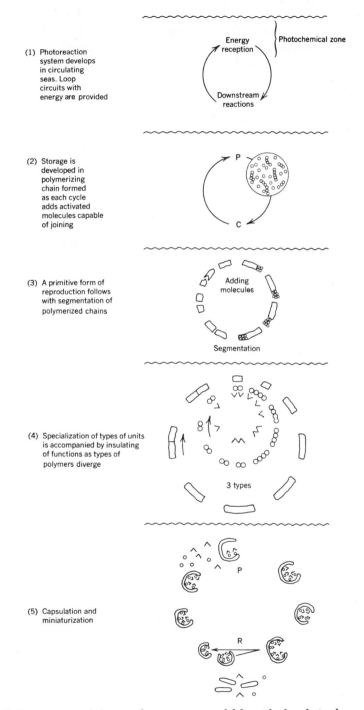

(1) Photoreaction system develops in circulating seas. Loop circuits with energy are provided

(2) Storage is developed in polymerizing chain formed as each cycle adds activated molecules capable of joining

(3) A primitive form of reproduction follows with segmentation of polymerized chains

(4) Specialization of types of units is accompanied by insulating of functions as types of polymers diverge

(5) Capsulation and miniaturization

Figure 5–7   Diagram of the step-by-step origin of life with the choice-loop-selector acting on circulating molecules energized with photochemical reactions.

minerals of the rivers and sea, plus the organic molecules that exist in such fluids, receive radiational energies causing oxidation-reduction separations and providing the opportunity for subsequent reaction. While cycling with the pulses of energy and reaction, the molecules may go from polymers that survive to polymers that lengthen and break, thus setting up organic reproduction. Circulating large molecules may reproduce with specialization, then with grouping, and then with increasing complexities and miniaturization toward the units we know as life. There probably never was an exact moment life started to exist, for the energetics and cycles were already there. The specialization, capsulization, and subdivisions into living units occurred step by step.

The energy diagram for the origin of life scheme in Figure 5–7 is the same as that in the modern biosphere (Fig. 1–5(e)). Comparing it with Figure 5–6, we recognize it as another example of a choice-loop-selector. The photochemical action is a random-choice generator; the circulating fluid provides a reward loop to choose for further use those units capable of the stages of evolution shown. Increased order is paid for energetically in small increments.

MINIATURIZATION

Many mechanical and electronic systems that are large and clumsy initially later evolve into miniaturized, economical systems. Engines, radios, and computers have gone through such stages of development. Miniaturization provides more function for less power and less space, ultimately allowing more function and complexity for the same resources.

We may visualize a similar possibility in biological evolution. Many groups of animals and plants first appear in the fossil record as large simple structures and are later replaced by smaller, possibly more efficient, species. There were at first, for example, large armored fishes, large ferns, trees, birds, dinosaurs, giant foraminifera, insects, and so on.

If the origin of life was a step-by-step capsulation of the loose circulations of molecules in the sea as suggested in Figure 5–7, the earliest forms must have been so large and had so little structure that they left little in the fossil record. Gradually, with miniaturization, some component structures must have become recognizable after millions of years of P and R circulation. Such a history would help explain the sudden appearance of advanced fossils in the Paleozoic rocks. Life from earlier periods may have been too large and loose to be recognized.

If early life formed in this way, with large loops preceding the small capsules, some of our best miniaturized modern world dominants such as the bacteria may not have been the earliest forms.

SPECIATION AND INSULATION

In Chapter 3 we introduced the concept of species as functional and specialized energy flows out of which complex networks were self-designed. The formation of species has been called microevolution, and an extensive literature shows examples of species generation. As summarized by such leaders as Mayr [9], a population that becomes divided by space or other factors generates genetic differences by mutations and genetic recombination phenomena. Differences in stocks are then exposed to differences in natural selection so that populations develop dissimilarities. In the language of the previous paragraphs, speciation like other creative work involves a random-choice generating process followed by a choosing system.

New species are formed when new adaptations receive better loop reinforcement and reward from the system in which speciation is occurring. For reinforcement the direction of the adaptive change must be toward greater energy flow through the population circuit. Adaptations that tend to circumvent system-limiting factors and increase flow are rewarded by the feedback loops. In this way species are developed for the compatibility of systems. The systems view regards concurrent speciation as a coordinated adaptation to evolving networks, but much of the literature has regarded speciation as a series of independent population processes. Most of the published research has not recognized the extent of network organization and loops as a speciation guide. Speciation is best considered as the process of adapting parts (populations) to evolving systems. In human affairs the evolution of specialized industries with their specialized occupational populations is an equivalent process; each industry receives its selective reinforcement according to its economic reward for contribution to other circuits. Some population biologists who deny system selection would, if they were consistent, claim that the industries most likely to survive are those that produce the most, regardless of market. There are markets in all ecosystems which feed back circular controls to the species.

The development of specific behavioral and chemical means for keeping the species functions separate constitutes insulation of the circuits and is expensive in work costs to the energy budget. If insulating mechanisms are absent, energies are lost through leakages between circuits.

The energy reinforcement for developing species-insulating mechanisms is greater when networks are complex. See, for example, the energy network diagram for a section of a rain forest (Fig. 3–11).

THE MALTHUS CURVE AND BIRTH CONTROL

Almost every one has heard of the Malthusian theory which predicts that if reproduction continues at a constant rate per individual, the acceleration will cause the food resources to be exhausted and the population to be limited by poverty, malnutrition, and disease. That the world population seems to be on such a Malthusian curve is widely

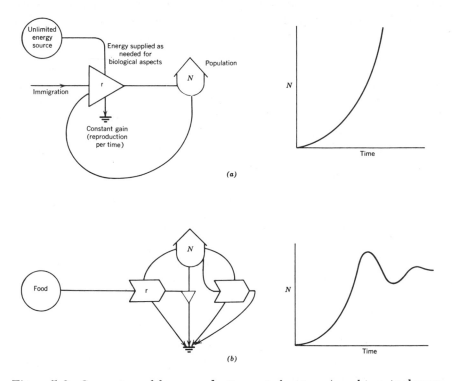

Figure 5–8 Comparison of four reproduction control systems (usual in animal populations and primitive man). (*a*) Reproduction where gain (per capita reproduction rate) is not energy-limited (modern man in industrialized culture); (*b*) reproduction depends on food; there are no other energy sources; growth is limited by crowding effects; (*c*) reproduction controlled by an energy sensor; (*d*) reproduction has a special control circuit with a prior energy storage for education which controls reproduction.

known. An energy diagram for this kind of growth is given in Figure 5–8(a).

When populations of microbes or animals in a laboratory are allowed to continue unlimited reproduction, crowding develops and with it many negative effects that are caused by the density such as interactions of wastes and behavior that resist the reproductive process. Mathematically these interactions are proportional to the square of the density. As shown with the model in Figure 5–8(b) the population ceases growth and may decline again.

If the energy source is limited, the growth will stop and the population will level off until the population demand equals the supply rate. If, however, the level of energy supply varies as it often does with season, there follows a lag, the population being at first too small for the energy supply and then too large.

In recent years many scientific works on the populations of organisms in nature have stressed the wide prevalence of a mechanism that prevents the reproduction per individual from remaining constant until the food supply has been exhausted. Instead the mechanism couples reproduction

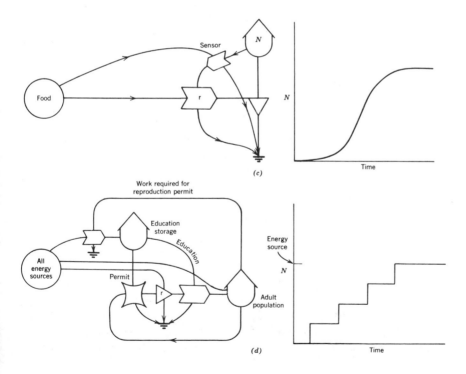

(c)

(d)

to the food supply and regulates it in proportion to the already detectable shrinkages in energy input as in Figure 5–8(c). Thus, birds cut down the number of eggs laid when food levels are low [7]; rotifers put out fewer eggs when algal levels are lower [5]. In short, many and perhaps nearly all natural populations in undisturbed systems practice a birth control [14] in which reproduction is in proportion to the energy budget. This system is, however, not only birth control but a means of encouraging birth when more innovations through evolutionary selection may be useful and energetically permissible. The energy diagram for sensor-controlled, food-limited populations is given in Figure 5–8(c).

When most of the energy available to man's system came through his food, the energy limit to the physiological system served to curtail reproduction immediately, in periods of poor food supply. Now the energies available to the system of man are in great excess of his food, and priorities of welfare go to adequate nutrition. Consequently, no physiological limiting stress develops even when energy is inadequate for other support and education. There may not be enough calories to cover college education, but the physiological system suffers no stress, for all the necessary food fuels are available. Figure 5–7(a) shows the system of man when food is always adequate; reproduction is at a constant rate per person. A triangular constant-gain symbol is used because the process of reproduction with energy available as needed is the same energetically as amplification of a voltage in electricity or broadcasting. The growth curve is Malthusian as shown.

The progress of world birth control now being undertaken lacks an automatic energy sensor for quantitative control which provides flexibility and stability to natural populations. For man's society some regularly calculated statistics on power input should be used as the means for determining the number of offspring that the system generates in any particular year. Since our energy supplies are still expanding, it may be that the system of man still has an energy sensor operating, but that it is indicating unlimited growth because of unlimited expansion in input energies.

One suggestion for human society is diagramed in Figure 5–8(d), using educational financing as an energy sensor. Reproduction is permitted only after work is done to guarantee the education of the new offspring. More reproduction is permitted when the reserves set aside for education are adequate. When energies for education are less, reproduction is cut off even though energies for the physical aspect of reproduction and childhood nutrition are adequate. To make the model work in a way that is consistent with democratic principles will require innovations in social institutions.

The criterion for natural selection has often been considered simply as the maximum possible reproduction and survival of the species concerned, a principle which would, if it operated alone, make shambles of any system of many species through destructive competition. Actually, simple mechanisms exist for connecting the many species into a single system with selection according to the harmony of the separate roles often serving to limit reproduction and survival. Any feedback from downstream that serves as a gate and selective control on an upstream population acts as such a mechanism. Lotka's principle of selection (Chap. 2) for maximum power output refers to all the useful power that flows into useful work as well as into storage and reproduction.

POWER AND MEDICAL MAINTENANCE

With billions of dollars spent for medicine each year, man seems to be setting out to eliminate from his own system the very principle of extra duplication, selection, and system loopback that keeps the order, eliminates the errors, and encourages evolutionary progress. His brilliant achievements of modern medicine are directed at eliminating the extra duplicating and at saving all the units duplicated for further replication. He has a loopback currency, but does it reflect the needs of the system in its stimulus to the duplicators? As the genetic errors are retained, medicine steps in with ever-greater programs and power requirements to repair, support, and permit the continuance of the errors. Compare the systems for control of age in Figure 5–9. In Figure 5–9(a) is a controlled carnivore animal system; in Figure 5–9(b) is primitive human system disease controlled; Figure 5–9(c) is our modern system.

In exchange for the accelerating cost of power flows that must go for maintenance of what was earlier not maintained, the system economizes by not overduplicating and losing people to early deaths and nonuseful lives. In effect, the old system for maintaining order (1 in Figure 5–9 (b)) is being replaced by man-made ways of repairing order and supplying services from the special energy budgets. Is one way any more humanitarian than the other? Does one have any energetic advantage? If the power requirements diminish, what kind of disintegration occurs in the medical maintenance system? These are open, sensible questions that quantitative data can help answer. How cruel runs our delusion, however, that all this represents great advancement when it is merely sub-

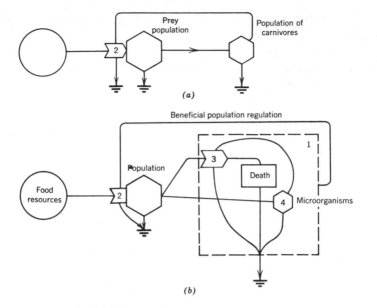

(a)

(b)

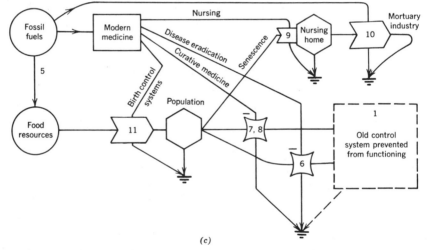

(c)

Figure 5–9 Comparison of energy circuits for management of aging and death. (*a*) Animal population with older individuals taken by carnivores. (*b*) Pattern for primitive man in which the population removal system is a disease organ-weakness complex (1) that derives its energy from the population; (2) beneficial control action; (3) energy stress on the individual that exceeds the energy reserve for supporting essential functions of key organs; (4) diseases that are the source of energy stress at point 3. The diseased population can be considered supported by the energy drained from the general population by carriers and mildly ill individuals. The removal system as a whole (point 1) serves as a useful selector maintaining the popu-

stitution of one cost for another, the modern system not having the billion-year test of evolutionary time.

QUALITY-CONTROL CYCLE RELAYS AND OLD AGE

The natural systems have means for removing the living parts when they begin to falter because of their accumulations of internal disorder. Thus the plankton of the sea and the antelope of the plains are consumed by carnivores which take the faltering minerals back into the system, regenerate them, and keep the chain of function at high efficiency. We can bring such plankton into the laboratory and omit the special-cycle species so that the plankton reach old age. Slobodkin and Richman [12] did this and found both symptoms of senescence of the individuals and lower power flows of the population, privations which would in the natural environment lose them their competitive position. The cost of maintaining the effectiveness of the chain is the cost of the carnivores, although the energy of the units eaten may pay for the quality-control function (see Figure 5-9(a)).

When carnivores or diseases are exerting an age distribution control, they are doing work beneficial to their prey populations (2 in Fig. 5-9) as well as drawing energy from them. Thus an energy diagram showed a work loopback in managing the cattle (Figures 4-2 and 4-3).

Man's national programs with their strange marriage of ethics and science are rapidly removing the aspects of his species that served as his quality-control relays. At an earlier time the human body maintained some weak points (3 in Fig. 5-9) and some parasites and diseases (4 in Fig. 5-9) which served neatly and cleanly to mineralize individuals and take them out of circuit when general deterioration brought body power reserves below certain minima.[4] Heart attacks and disease and

---

[4] Strehler [13] used the biomathematics of death curves with time to show that human death may often be caused by the chance that the energy demands of random stresses exceed reserves as the efficiency of functions decline with senescence. He showed how to derive the Gompertz equation from this premise. Human death curves follow the Gompertz graph from age 30 to 75. A Gompertz curve is straight on a log-log-graph with time.

---

lation with an effective age structure. (c) Pattern for an industrial culture with modern medicine on fossil fuel subsidy: (5) augmented nutrition in places of shortage; (6) disease protection and prevention; (7) maintenance of individuals with formerly lethal defects; (8) elimination of the critical organ sensitivities with heart medicine and similar medications; (9) rest home care; (10) much delayed death and burial; (11) birth control efforts.

their interactions served to remove individuals just as the carnivores eliminated the old plankters.

Now, however, magnificent modern medicine using a thousand preventive means (6 in Fig. 5–9), including nutrition, immunization, disease control, and more recently work toward the elimination of early heart failures, carries the individual far past these stages.

Curative medicine is even more effective in reducing early deaths (7 in Fig. 5–9(c)). Much energy goes into mortuary business also [10]. Having eliminated the disease control system producing a population explosion, we are forced to spend more calories in making a new birth control system (11 in Fig. 5–9(c)).

The emerging pattern is one of the great industrialized nations putting huge sums into human tissue culture—that of ourselves as great-great-grandparents (9 in Fig. 5–9(c)). What a wonder it is that the new youngsters can visit their forefathers or at least some clump of vegetating cells still thus named in one of the mushrooming new nursing homes. Is this the continuation of great people, or have they already relinquished their souls in the information they have passed forward in their offspring, in their teachings, in their writings, and in their contributions to the continuation and evolution of the great societal networks? We might wonder whether disease and subsequent death served primitive man well, both for himself and for the effectiveness of his system role. No doubt increasingly vast power flows will go into the vegetable cultures. No doubt if there is retrenchment of the rich power sources, the short-lived system will return with a vengeance; but as long as we have some primitive, disease-resistant stocks available from underdeveloped countries, man may survive. This paragraph has stated the negative face of our medical ethic.

THE CONSERVATION OF INFORMATION STORAGE

There may be another, more positive, face. With the increasing complexity of the power-rich networks, the role of the human computer (man's educated brain) becomes ever more valuable with greater and greater power investments in its training and memory storages. Now individuals begin to be truly unique because combinations of knowledge and training allow them to be and do things no one has ever done or may ever do again. The value of these individuals who have received educational investment is vast. Perhaps the power cost of sustaining them well into the vegetable stage is justified.

There was a time when hordes of young adults could wield their brash and dash, conquer, and run the world. How long now does it take to learn enough to pull one's weight and develop a loop reward in the complex society? The dash and brash now goes into street behavior and into nonproductive subcultures marking time. Empty computers are not needed.

IS THE SELF-DESIGNING PROCESS OPERATING TOWARD SURVIVAL?

From the example of modern medicine, it is clear that the vast new energies from fossil fuels are producing extremely rapid evolution of the modern human system, displacing and sometimes actively preventing the operation of the older successful systems of human continuance and control. Although the seething ferment of trial and error in the experiments with social institutions constitutes an active generator, and economic forces at times supply loop reinforcement for some innovations, are overall controls being evolved too? In part, overall control is still being exerted by old religious guidelines. Are the necessary overall control systems evolving as fast and in the right direction to keep the overall system viable?

NATURAL SELECTIONS AND EVOLUTION IN NONLIVING SYSTEMS

Where power supplies are available, living systems develop systems rapidly capable of drawing power competitively and they build whatever structure of species is necessary to process the energies. As we have already described, the selection of components from a choice is the means by which the pathways of greater power delivery in loop reward actions are reinforced, rewarded, and retained.

The same processes operate in nonliving systems. Whenever there are substantial power supplies available, a selection from among eddies and other energy flow choices rewards and reinforces those systems that develop larger circular loops that reward the pathways that draw the greater power. In this way potential energy differences develop circulations in atmospheric convection, weather systems, and hurricanes. Figure 5–10 is a diagram of energy flow in a weather system. At least in the abstraction of processing energy, circulating matter, and maintaining a structure while undergoing evolution, the weather systems have much in common with life. Many features of fluid dynamics may be explainable in essentially biological terms. The ratio of forces required to start

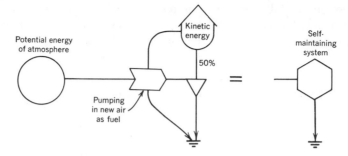

Figure 5–10   Energy diagram for hurricanes and other cyclones showing the similar patterns of self maintenance found in life.

eddies is described by Reynolds number for two adjacent flows at different speeds and by Richardsons number for one flow rising from a heat surface relative to another fluid. The existence of these numbers suggests that there are energy level thresholds. There should be energy levels at which there is enough choice and selection to set up recycling pumps, maximizing power flows. Even in the microscopic world of chemical actions, it is likely that systems that develop circulatory pumping activities override strictly chemical behavior.

INFORMATION, ITS SPECIAL SCIENTIFIC SENSE

Information to most people is a broad term covering words, data, messages, codes, and other inputs to the human mind which are stored in the memory or in libraries and which are the basis for effective actions. In recent years, a way to measure information from Shannon and Wiener has come into use which has a more restricted but quantitative definition. The amount of information is measured in two steps: first, measure the complexity of whatever is being examined (the message, system, configuration, coding, species association, molecular association, etc.); second, specify the complexity in the particular combination if it is known.

In the system in Figure 1–7, we can examine the number of possible pathways. A count of these is a measure of the complexity and also of the uncertainty inherent in a situation of this complexity. With 16 possible pathways, there are 65,536 possible network arrangements of which the one in Figure 1–7 is only one. Each pathway opportunity is like a coin in having two choices (connected or absent) requiring one decision. A "bit" is the unit of information[5] and is defined as the amount of uncertainty in the situation of one decision between two possibilities. For each possible pathway one adds one bit of information. There are 16 bits of information in a network organizing 4 units as in Figure 1–7 (16 decisions). Costs of description and organization may be proportional.

Describing any system like that in Fig. 1–7 also requires specification of the functional units, many of which are species populations, each with a different proportion of the total. For example, there may be 100 plant individuals and 10 animal individuals in the system of Fig. 1–7. There is uncertainty and thus information in the opportunity for various ratios in different species, quite apart from pathway organization which we have already discussed. Introduced by Margalef and widely used in ecology are studies of the information content associated with the species composition. Species information is calculated[6] as if the species were letters in a message, which they actually are in the message of inheritance by which ecosystems are maintained and developed. Along with graphical expressions of species diversity like those in Figure 4–11 the species information is used as a diversity index.

The amount of information increases with the number of units included in the system. A square mile of trees has more information than an acre. To measure the degree of concentration of the information we may divide the information calculated by the number of individual units involved. In computations of the information content of species combinations in ecosystems, values range up to 6 or more bits per individual because of the many possible combinations. The resulting number, although called information, does not indicate whether the complexity is organized into a useful combination or whether it is an unspecified

[5] Information (I) can be defined as the logarithm to the base 2 of the possible combinations (C). Thus $I = \log_2 C$. A logarithm to the base 2 is the exponent one uses to multiply 2 by itself to get C. In the example of Figure I–7, C is 65,536 and I is 16 if only one pathway is allowable for each box of the matrix shown there. When one combines two parts of a system, one may compute the total possible combinations in the new complex by multiplying the possible combinations in each part. Since logarithms are exponents, one accomplishes this by adding them. Information is thus an additive measure of complexity.

[6] Where a system has many kinds of items, each present in different ratios such as letters of the alphabet in a message or species in a forest, there is a form of the information formula which gives the bits per individual (H) due to the composition in the system or in information messages used to describe and transmit it. Thus

$$H = -\sum_{1}^{n} p_i \log_2 p_i$$ where p_i is the probability of each type of item in the system and n the number of kinds.

random situation. The information content calculated as the logarithm of the combinations indicates the amount of useful information you would have if it were organized into a useful message or it indicates the amount of jumble you have if it is not organized. We arrive at the same value for species thrown into a random association as for a tightly organized system. The ecosystem example (Figure 1–7) would provide the same amount of information, if the arrangement of pathways was a workable network or the random situation with no pathways usefully organized.

DETERMINED AND UNDETERMINED INFORMATION

The second step in the process of indicating the amount of useful information is to specify what part is a controlled combination and known to be organized. When determined the pattern is no longer uncertain, the information measure then indicates the amount of decisions and work necessary to arrange a system, to keep it organized, or to transmit the message to another place or to the future using memory.

The total information is the sum of that defined and recorded and that uncertain and unorganized.

$$I = I_{\text{organized and known}} + I_{\text{uncertain}} + I_{\text{known to be disorganized}}$$

For the example in Figure 1–7 only 8 pathways are actually organized representing 8 bits of information although the situation has 16 bits of uncertainty (in a system of one pathway between each combination of units). Subtracting we find 8 bits of unrealized organization.

INFORMATION CONTENT OF MOLECULAR PATTERNS

When the number of units includes the billions of molecules in a gram of matter, the information content becomes very large. A gram of water, for example, has about 3×10^{22} molecules which gives us an immense number of possible combinations in which to connect them. The information content is huge.

In this case most of this information is uncertainty, not organized by anyone or any process into a particular combination. The information content here measures only the disorder. There are far too many units to be completely ordered at this temperature because the heat energy (which is molecular) keeps the molecules vibrating and shuffling.

Earlier and independently from the Shannon-Weiner use of information the logarithm of the number of molecular configurations was found to be a measure of the disorder of chemical substances and, if multiplied by Boltzman's constant, identical with the measure: entropy.[7] In Chapter 3 we mentioned entropy as increasing as a measure of disorder of the energies which go from potential state to dispersed unavailable state. Entropy is information content of the disordered molecular states in chemical and heat units of measure.

If the water evaporates, the molecules move about more freely as vapor producing even more randomness than before, with more combinations of states possible. The entropy has increased. The original state did have some initial organization lost during the evaporation that increased its randomness.

POOR INFORMATION TRANSFER WITH HIGH-POWER FLOWS

As we indicated earlier power will "out" unless work is done to prevent it. Because of the tendency for potential energy to develop work circuits and generate much energy-expending routes such as eddies, noise, short circuits, and many other losses, the amount of random processes increases with the amount of power. Power lines, highways, factories, jet planes, and fast rivers are noisy. The word noise is often used, not only for sound but for all kinds of random variations and losses in energy flows.

If high-power flows are noisy and full of random variations, they are not the best media for sending low-level information messages. A power administrator does not put a telegraph key on his 100,000-volt power line transmissions to the next city when he sends a message; he sends his message over a 12-volt telephone line. Similarly in the ecosystem we see the advanced systems differentiating their power circuits and their information circuits although both are energy transmissions. The energy amplification value of one is high, but the actual power level of the other is high.

[7] Boltzman-Plank formula is

$$S = k \ln m$$

where m is the number of unorganized molecular states, k is Boltzman's constant, 3.3×10^{-27} kcal/(°C)(molecule). A mole is 6.02×10^{23} molecules. Here log to the natural base e is used (2.718). To convert units use the following: 1 cal/degree = 8×10^{23} bits. There is about 1 bit per molecule for each cal/(°C) (mole).

In some papers the definition of information and the definition that was independently made for molecules so as to yield entropy were done with different algebraic signs. Either is correct if used consistently. We can talk about positive and negative information and positive and negative entropy, without really changing anything except the agreed-upon convention. As disordering proceeds negative entropy increases in magnitude as does positive entropy. They both would become zero when all disorder is converted into crystalline order at absolute zero.

INFORMATION PATHWAYS

The transmission of information is an important part of any complex system. A plant manager makes his company respond on the basis of a stock market report. A cell makes its biochemical machinery respond on the basis of codes received from its genes. An ecosystem makes its power flows respond on the basis of its memory storage, some of which are biological and some of which may be physical or in libraries, in records, in rocks, or in wood structure.

Specified information even of small messages of 3 bits on paper has with it a small content of potential energy since its position is less probable than would develop at random (Brillouin [2]). The normal disordering that would occur if there were no maintenance would move from the ordered useful pattern to a random, unspecified one. Potential energy would have passed into disorder (heat). For such situations of a few combinations found in messages, the energy content as a fuel is far too negligible to measure or consider compared to the great flows of energy in the food chain. Yet the quality of this information (tiny energies in the right form) is so high that in the right control circuit it may obtain huge amplifications and control vast power flows.

Information pathways, although of low energy, are still energy flows and may be shown on the energy diagrams along with the pathways of higher power. See, for example, Figure 7–2 where human opinions are transmitted, stored, restated, and used to control power actions. In the transfer of information, there must be some kind of carrier power flow such as the electricity in the telephone line or the energy of a flying bird carrying seeds.

INFORMATION VALUE

Small energy flows that have high amplification factors have value in proportion to the energies they control. As the smallest of energy flows information pathways may have the highest values of all when they open work-gate valves on power circuits.

The information explosion is a perplexing property of public decision, for great costs are involved in storage of information. When is it cheaper to develop information and when is it cheaper to store and relocate it? In part it has to do with the energy value of the information in its use. The value of our information decreases as it becomes more difficult to find it in our own information complexity. Communication engineers describe limits in transmitting messages by the capacity of the channel and the ratio of signal to noise. We have a noisy channel of information transfer from the past to the future. Can we put too many messages in it?

REFERENCES

[1] Barnes, H., and H. T. Powell, "The Development, General Morphology and Subsequent Elimination of Barnacle Populations *Balanus crenatus* and *B. balanoides,* after a Heavy Initial Settlement," *J. Anim. Ecol.,* 19, 175–179 (1950). Sketch of original photograph from Kendigh, S. C., *Animal Ecology,* Prentice Hall, 1961.

[2] Brillouin, L., *Science and Information Theory,* 2nd ed., Academic Press, N.Y. (1962)

[3] Coon, C. S., *The Story of Man,* Knopf, New York, 1954.

[4] Copeland, B. J., and H. D. Hoese (1967), "Growth and Mortality of the American Oyster *Crassostrea virginica* in High Salinity Shallow Bays," *Publ. Inst. Mar. Sci.,* 11, 149–158 (1967).

[5] Edmondson, W. T., "Reproductive Rate of Planktonic Rotifers as related to Food and Temperature in Nature," *Ecol. Monogr.,* 35, 61–111 (1965).

[6] Goss, R. J., *Adaptive Growth,* Logos Press, Academic Press, New York, 1964.

[7] Lack, D., *The Natural Regulation of Animal Numbers,* Clarendon Press, Oxford, 1954.

[8] Leopold, L. B., and W. B. Langbein, *The Concept of Entropy in Landscape Evolution,* U. S. Geol. Surv. Paper 500A, 1962.

[9] Mayr, E., *Animal Species and Evolution,* Harvard University Press, Cambridge, Massachusetts, 1963.

[10] Nilsson, N. J., *Learning Machines,* McGraw-Hill, New York, 1965.

[11] Schrödinger, E., *What is Life?,* Cambridge Univ. Press, Cambridge, 1944.

[12] Slobodkin, L. B., and S. Richman, "The Effect of Removal of Fixed Percentages of the New Born on Size and Variability in Populations of *Daphnia pulicaria* (Forbes)," *Limnol. Oceanog.,* 1(3), 209–237 (1956).

[13] Strehler, B. L., "Fluctuating Energy Demands as Determinations of the Health Process in Biology of Aging," B. L. Strehler (ed.), *American Institute of Biological Sciences,* 1960, pp. 309–315.

[14] Wynne-Edwards, V. C., *Animal Dispersion in Relation to Social Behavior,* Hafner Publishing Co., New York, 1962.

6

POWER AND ECONOMICS

MONEY AS A LOOP REWARD SELECTOR

When we talk of the budgets of business and the affairs of man's systems, we are accustomed to the concept of money and its transference in exchange for goods and services.

Money is a special currency evolved to allow the production of one person to be rewarded by a feedback loop from some other part of society. Some authors have compared money to energy, but the two are not the same and they flow in opposite directions. We receive food from the grocery store by passing money in the opposite direction to the grocer (Fig. 6–1(a)). We receive money when we put energy into work (Fig. 6–1(b)) that makes an energetic contribution to the function of at least one other unit. The money system keeps its members contributing useful work for the network and money provides a means for organizing energy flows. Money circulates whereas energy flows are unidirectional (see the simple example in Fig. 6–2(c)). Let us trace the energy circuits for economic systems as they evolved from the simple currency loop to barter and payment and then to money.[1]

NATURAL CURRENCY

The living forests and reefs have an economy, but instead of money the currency is made up of necessary materials like phosphorus, nitrogen,

[1] Boulding [1] draws many comparisons between ecological systems without man and human economies, including a comparison of currency to mineral cycles. The energy network language may help formalize quantitative comparison of many of his points. Representative of the trend for the use of terminology, Boulding, an economist, suggests "ecosystem" as a suitable name to include economic systems as well as other aspects of man and nature.

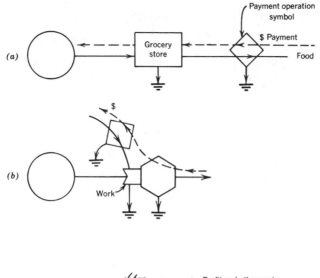

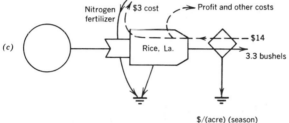

Figure 6–1 Relationship of energy flow to money flow in an economic transaction. (a) A process involving transfer of potential energy (food or fuel); (b) a useful work service in an effective production process of industry; (c) nitrogen fertilizer stimulation of rice in Louisiana [4].

potassium, carbon, water, and the exchange of work services. Plants use the minerals to make food, and then consumers eat the food and send their wastes back to the plants. The dashed lines in Figure 6–2(d) show these nutrients circulating as a currency. Any compatible combinations of plants and animals are self-rewarding, as we have discussed in Chapter 1 (Figs. 1–1(b) and 1–5(d)). Whenever there is no loopback of minerals from wastes to the plants, a necessary function for both animals and plants is interrupted and these species drop out. Whenever the rewarding loop is reinforced, that circuit continues. Through such loops and the intertwining of loop circuits the system has a means for selecting only the species and circuits that contribute to the whole functional cycling,

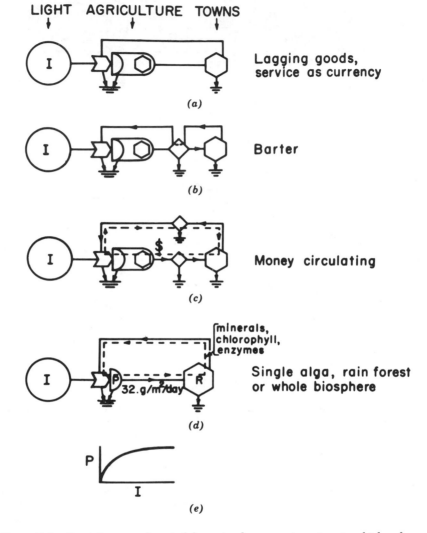

Figure 6–2 Several stages of control for a simple economic system in which a farmer makes food on a small plot and supplies it to a friend who works in the fields. Work flow aids the farm work by supplying labor, equipment, and fertilizer. (*a*) This simple loop system has no formal payment operations; (*b*) a more formal control, a barter payment control procedure has been added; (*c*) a more flexible control system has been added by circulating currency as a feedback payment; (*d*) cycle of materials in ecological and biological systems moves in the same direction as the energies; (*e*) short-term response of cycle limited systems to variations in input power.

in the same way that money is paid only for useful work. In electronic systems electrons circulate as the currency whereas their concentration determines the energy storage and driving voltage.

NUTRIENT RATIOS AND THE INPUT-OUTPUT MATRIX

An approach applied in recent years to studies of the economy is analysis of the input-output matrix. Each industry in a network of industries requires several inputs of material and labor in definite ratios to produce the final product. If inputs are available in smaller quantities the process is limited, and if an input comes in excess it is waste which piles up and requires special attention. With many industries in a system, each with its special requirements, the inputs and outputs of each are related by their necessary exchanges. The whole system can be limited through poor coordination and short supplies of particular components.

Figure 1–7 is an input-output matrix for carbon flows among 4 compartments of an ecosystem. Figure 1–7(c) is an analogous economy of four compartments with money circulating in reverse direction from the materials and services. Similar networks exist for the nitrogen, phosphorus, and other currencies of the two economies. In our economy man serves at the system control level as a selector of the various possible combinations in which the processes may be arranged. In actual operation feedbacks of more dollars paid for something in short supply serve as an automatic reward to the needed components so that they become available and the whole system tends toward the optimum.

The reader will recognize in the economic system loopbacks similar to the minerals and work processes discussed earlier for ecological systems. Redfield's principle (Chap. 1) about conservation of nutrient ratios in closed mineral cycles of oceanic plankton is the same as the principle of complementary flows of input-output networks in economics. Compare the two diagrams in Figure 1–7 which have the same input-output matrix. We readily recognize that the theory and practice for study and management of industrial networks and ecological networks are conceptually the same, except that the common denominator used in economics is money which is not used in the nonhuman ecosystems.

LINEAR PROGRAMMING

Where there are two or more units contributing to two or more processes, the system may be analyzed by a mathematical approach called linear programming in order to determine the optimum ratios of the units. For example, in Figure 6–3 there are two parallel agricultural industries consuming phosphorus and nitrogen fertilizer. Each crop has its characteristic ratio of utilization of the two nutrients and the system of fertilizer supply is providing them in an N/P ratio of 7 : 1. The model given in Figure 6–3 also applies to two species of phytoplankton algae in the sea each drawing from the nutrient supply upwelling from deepwater reserves in a ratio of 7 : 1.

The effects of each unit on each process are expressed by a term in the two equations, one for phosphorus and one for nitrogen (see figure legend). Each line is plotted on a graph that relates the two units (W_1 and W_2). Where the two lines intersect in Figure 6–3, utilization of the incoming fertilizer nutrients is complete without wastes or shortages. Other criteria being the same, selection by the closed loop of recycling nutrients may adjust the ratio of weights of W_1 and W_2 as indicated.

In an ecological system of an ant colony, Wilson [5] applied linear programming ideas and suggested that each caste of ants (such as soldiers, workers, etc.) acted as an industry in the economy of developing queens for further reproduction. Towards this ultimate economy several tasks were identified such as food preparation, defense, and nest building. The shifting in ratios of castes and the number of castes was related to the optimum group contribution of the specialists and processes. Linear programming thus helps us see how the ratios of specialists among the subsystems may be optimized. By these examples we find the small ecosystems of nature similar to those of the world of man, both processing raw materials and products, both having specialized subsystems, and both having criteria for optimization.

SYSTEMS WITH A BARTER PAYMENT CONTROL

In the examples of man in nature examined in Chapter 4 we observed the loopback of work from the downstream consumer to the upstream unit. Another example is a farmer supporting a friend with food in exchange for work by that friend (Fig. 6–2(a)). The positive feedback reward system becomes an organized economic barter system when the downstream and upstream feedback flows are linked and controlled by agreement so that one is made in proportion to the other at a point between units, each flow serving as a payment for the other. The payment transaction takes a small amount of potential energy from both ends and spends it in such things as transport and negotiations. In the energy diagram the control function is indicated by a diamond operator symbol

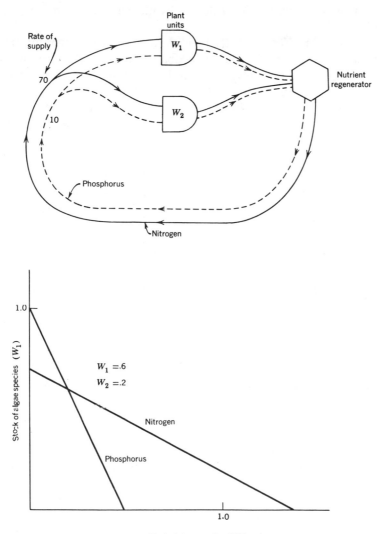

Figure 6–3 Linear programming of a system of two producing plant units (W_1 and W_2) each with characteristic rates of utilization of phosphorus and of nitrogen. Rate of supply of the fertilizer units is 7:1. Equations for use of 10 units of phosphorus and 70 units of nitrogen inflowing per day are: phosphorus: $10 \ W_1 + 20 \ W_2 = 10$; nitrogen: $100 \ W_1 + 50 \ W_2 = 70$. If selection works to adjust the ratios of W_1 and W_2 so that total utilization leaves no accumulations or shortages, then an optimum is located by the intersection of the lines. Reasonable figures for plankton are in mg/ (liter).

179

and a small heat drain. We call this work control process a *payment operation.*

The system in Figure 6–2(*a*), with no payment operator, is self-regulated by means of the closed feedback loop, but if there are appreciable energy storages, the system will have a time lag and may oscillate (Chap. 2). Rewards for balanced input-output relations require slow selective processes to establish patterns. By introducing the payment operation, the upstream and downstream flows are linked, and neither process can get ahead of the other. The payment operator also serves as a sequentially operating logic circuit which may be stated as follows: the systems in Figure 6–2 are all limited upstream by rates of supply of downstream quantities that act on the first work gate. As in Figure 2–6(*b*) this pattern leads to overall hyperbolic short-term responses to varying power (Fig. 6–2(*e*)). If a unit of currency flows, a unit of power or work flows.

The invention of some type of circulating symbolism allows more flexibility in the use of a payment control operation. As shown in Figure 6–2(*b*) adding money doubles the number of loops and provides a common denominator representing work flow. In Figure 6–4 the system is expanded to include taxes and government. The availability of money for payment at control points eliminates the need for complex barter contracts. The use of money as a common denominator provides rapid loop reward reinforcement from the system in proportion to work done for any part. The money allows flexibility in switching the circuit network because rewards are interchangeable among circuits.

For many purposes it is unnecessary to draw the money pathways on the diagrams, since they are parallel and backward from the energy flows and can be adequately represented with the operator symbols (Fig. 6–2(*e*)). Some symbolic economic systems may exist among animals. The species of rat that substitutes objects for items it takes from man is a curious example; the consequences of his act are unknown.

BALANCE OF WORK PAYMENTS

If systems are to operate without shortages and accumulations there must be a balance of circulating quantities such as minerals and money. The phrase "balance of payments" is a common one in public discussions. Although in Chapter 2 we emphasized that potential energies are dispersed into heat sinks diminishing downstream, we also showed the amplification of the feedback by which it regained its value to the cycle. When this criterion is achieved, the reward loops balance upstream and

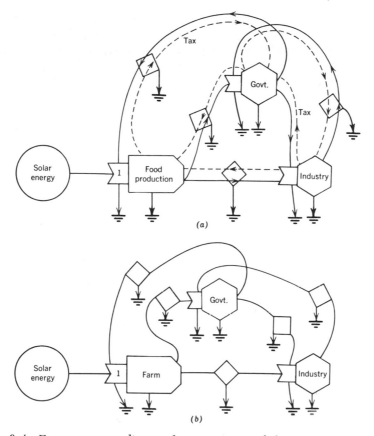

Figure 6–4 Energy-currency diagram for an economy of three compartments, agriculture-food production, industry and cities, and government. (*a*) Energy diagram with money circuits shown; (*b*) abbreviated energy-money diagram.

downstream work payments. The flows of energies like those of materials and money thus have a balance of flows made possible by the amplification of input energy sources. In effect, a system is stable when the amplification factor of the feedback loop equals the potential energy losses along the circuit. The upstream units are adequately paid for the downstream drains of the units there when the feedbacks through amplification restore the energies of their drain.

In Chapter 4 the energies spent in work developing a quality product, services, or some information, were used as a measure of the value of that item. If a closed loop exists of the type in Figures 1–5(*e*) and 6–2(*a*), the work spent on the downstream flow measures the replacement energy value of the downstream unit. The value of that unit's work upstream is measured by its amplified energy upstream. In a closed loop the energy of the return loop amplified must equal the energies spent downstream. Thus in a closed loop the two energy measures of value are equal. The replacement value and the amplification value become the same. For example, in Figure 1–5(d) 3.9 kcal/(m^2)(day) are used in developing the high quality of the downstream respiratory operation which in feeding back to the upstream photosynthetic process through its control gate action regains 3.9 kcal/(m^2)(day)—a measure of its value as amplification.

MONEY EQUIVALENT OF WORK

In traditional economics as illustrated by Figure 6–2(c), an input power flow has no money equivalent if it is free. When some energy source enters man's economic system apparently without charge, as when the sun falls on crops, wind turns the farm windmill, the river turns the waterwheel, or coal is gathered, the money value of the energy is in proportion to the work spent in receiving the energy and not in the energy itself. Oil costs reflect the distance the oil must be transported from its origin and the difficulties of pumping it, rather than its energy value. Whenever work is done by people, dollar value develops.

Although potential energy has no dollar value, there is an average dollar value equivalent to the work done in the economy. There is an average ratio of the money circulating in counter-current against the work services as drawn in Figure 6–2(c). The ratio may change, however. For the United States Gambel [2] gives ratios of energy consumption to income (Fig. 6–5) and shows the ratio for different countries (Fig. 6–6). These are not total energies since the photosynthetic system's reception of light is not included. The ratio of fossil-fuel energy to dollars has been decreasing. The ratios range between 7000 and 30,000 kcal/dollar (about 4 Btu/kcal). Elsewhere in some calculations in this book, 10,000 kcal/dollar was used. If energy per person is a measure of the standard of living due to fossil fuels, it can be seen in Figure 6–5 that there has been little change even though the energy levels and total economy are accelerating very rapidly.

The science of economics may profit by restating more of its theorems to include power principles. Studies of money alone are just as incomplete as studies of mineral cycles alone. Both consider pathways and flow rates without examining the driving forces that are generated from the

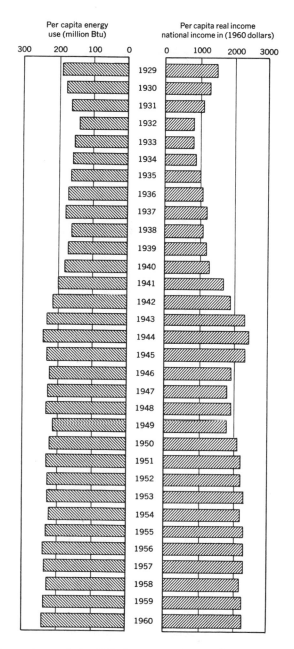

Figure 6–5 Comparison of fossil-fuel consumption and incomes in the United States [2].

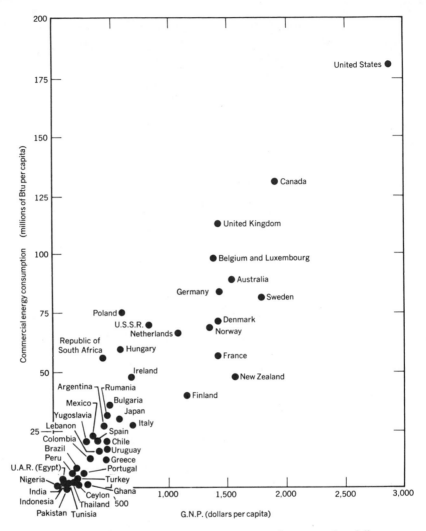

Figure 6–6 Comparison of fossil-fuel consumption and incomes for different countries [2].

potential energy distributions. Can predictive laws be formulated without using forces and energies?

One useful aspect of a dollar-work equivalent computation is that it provides a means for comparing the economical considerations of one time period with those of another. Does the energy spent in bringing in new energy affect the dollar-work ratio?

THE SHIFT OF MAN'S ROLE DOWN THE ENERGY CHAIN

As fossil fuels are injected, the role of machines increases, outcompeting man in simple, mechanical work. The increased total work done increases the standard of living but only to those who can plug into the economy with a service that has an amplification value greater than the machines. These pressures are forcing the human population to develop higher and higher educational levels so that their services can be valuable enough, when amplified, to obtain for them their prorated share of the economy. The problems of poverty become greater for those who do not have the means or motivations to move down the long educational path.

As the machines get more sophisticated, man is forced out of labor and out of blue-collar, mass-production work. Man seems secure from competition in the mental pursuits in which the human brain with its magnificent miniaturization may excel. These aspects show up in energy-currency model diagrams (Fig. 6–2(a)). Compare the circulating currency, which is the same in each part of its loop, with the energy flow in counter-current coupling to the currency. The energy decreases downstream but with increasing ratio of currency to energy. Men who can move further down this chain and gain greater amplified action increase their incomes. Then with many people with this amplification ability, the system's power is increased to make opportunities for so much intelligence. Thus the situation worsens with a rising Malthusian power surge.

ECONOMIC CONCEPTS IN THE ENERGY DIAGRAM

The energy diagrams allows some of the common concepts of economics which are also meaningful in ecological systems to be stated in the energy network language. In reciprocal manner the concepts of ecology usually expressed in energetic language may be identified with the economic terms. Consider some examples, where concepts may be similar.

A *good* is an accumulated result of useful work. (Its *value* is the time integral of the flow of potential energy expended in work.) An object or a service requires accumulated work to bring it into existence, starting from a particular point in this system.

A *want* is an input circuit upon which a downstream compartment is expending work to pump in the item wanted (see work flow 3 in Fig. 6–7). The energy for the input pumping may come from a source other than the recipient (flow 2 in Fig. 6–7) or even from the upstream source (Fig. 6–7, flow 1).

Production is the flow of goods, materials, and energy generated by

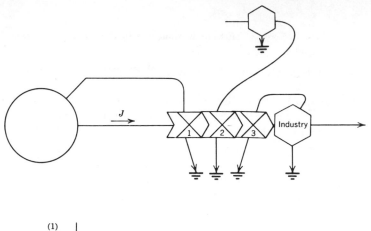

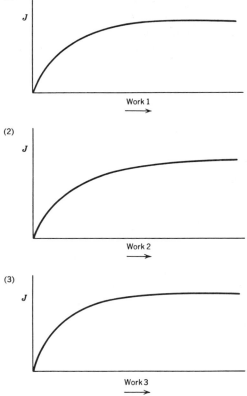

Figure 6–7 Curves illustrating the response of the main energy flow to increases in the work flows, one at a time, each in turn serving as a limiting factor (law of diminishing returns). The work flows marked with X indicate a multiplicative effect, as with the pumping in of required input components. (1) Work flow from an external energy source such as the sun on rain; (2) work flow from another network unit; (3) loopback work from a unit to its own process (such as reuse of a material).

one or more work flows on the circuit. It is the flow to the right from the agriculture symbol in Figure 6–2, for example.

Cost is the potential energy flow into heat necessary for a process. For systems with an economic payment it is also measured by the proportional money flow through the payment operator in a direction opposite to that of the energy flow (Fig. 6–3(*d*)).

Investment is the potential energy expenditure into heat and corresponding flow of money through the payment operator when the work done increases the amount of structure and order of a unit of the system's network. It is equivalent to the reproduction-maintenance-growth flows within the hexagonal symbol (see Fig. 3–13).

Income to a unit is the input of potential energy capable of doing work in that unit and thus earning money for that work. Income of money and income of fuel energy are thus related and are proportional at steady state.

A *firm* is a compartmental unit of the system as analyzed in an energy network.

What is called *equilibrium* or *dynamic equilibrium* in economics is best called steady-state flow in order not to confuse it with the force-equal equilibrium of reversible states that have no energy transformation.

Economic growth is an increase in the size of a unit of the system and/or its flows. It may be expressed in units of energy storage or of energy flows, in the money that flows through the payment operators in proportion to energy storage or flows, or in units of value as defined earlier.

Energetically, *inputs* are the sum of necessary work processes required to bring the necessary materials and energy flows together in the combining ratios (reaction ratios) of the outputs.

The *law of diminishing returns* of economics turns out to be the limiting factor relationship of plant physiology, agriculture, and population ecology, where the several flows of necessary input have a multiplicative action on the input flow. As any one factor is increased, it gradually passes through its limiting range to a plateau as other members of the flow become limiting (Fig. 2–6(*b*)). Figure 6–7 is an example of the effect of increasing one multiplier flow. The equation, an hyperbola, is discussed in detail in the papers of Ikusima [3]. The mathematical formulation is known in different fields by different names: Michaelis-Menton equation in biochemistry, the Monod relation of microbiology, and the Lumry and Spikes relation of photosynthesis. See Appendix.

Purchasing power of money is the ratio of the dollars to the total work involved in the economy. The calorie contents do not change, but the number of dollars being circulated in relation to the work done does change, sometimes by the mere act of printing money or devaluating currency.

Competition in ecology and in economics refers to many separate but related phenomena. The energy network diagram helps to define each phenomenon clearly and to consider it separately. For example, there are parallel circuits from the same energy source. Such flows draining the same source may, if they are stabilized by some special controls (work spending), provide means for more network diversity and flexibility, but if they are not so stabilized one may act to eliminate the other through competitive drain of the resource. In another phenomenon some systems may spend work energy to close off one or more inputs to other parallel flows so that they themselves receive more input. Such systems may help themselves temporarily, but this aid is not necessarily loop-reinforced from downstream since it may create disharmony, accumulations, and limits elsewhere in the system. Such self-motivated, destructive systems may also induce switching circuits from government, from ethical controls, and so forth to be directed against them. These switching circuits limit and regulate the destructive systems and receive loop reinforcement for their constructive actions from network flows elsewhere.

Credit is the flow of work for which the money loopback is delayed. Such time lag can allow oscillations or can be used to dampen out oscillations in regions of sharp season. The ecological systems operate mainly on credit. For example, the plants in spring produce for the animals from their reserves of mineral currency, whereas the payment by animals and microbes of minerals loops back to the plants probably during the winter, long after the harvest but in time to start a new cycle.

Circuits of a system that have to be maintained but are not being used for system work can be described as *unemployment*. A certain part of any population needs to be unused to provide reserve capacity and to perform full-time maintenance functions and information-increasing actions such as sleep, education, and relaxation. However, too much unemployment means a system with too much maintenance cost. Unemployment can be diminished by increasing energy inputs in channels that allow more work or by decreasing population in relation to energy input.

The *distribution of incomes* is really a power spectrum of energy circuits and usually has a skewed pattern because fewer large energy flows are needed than smaller ones. There cannot be energy flows much below a minimum value for maintaining a circuit.

The action of *borrowing* money is a network switch that provides for a new energy flow and adds new system structure, thus tapping new potential energies and forming new circuits. A bad investment fails to develop a new energy source and thus is unable to pay back money representing work energy.

Interest is an increased return that provides the loopback to the source which originally loaned work, thus forming an energy reward loop to the loan source from the new circuit made possible by the new potential energies tapped. Ecological systems have flows equivalent to interest when one unit spends work to set up another, as in ordinary reproduction, various kinds of symbiosis, and the work done by pioneer plants for successors.

THE PRICE OF WORK AND GOODS

The money paid for work flow in the payment operator (Fig. 6–1) is the price in energy terms. In part, the price depends on the total amount of money in the system as related to the total amount of work being done by the individuals who comprise the currency system and accept its laws. The work done by individuals depends on the power flows they have at their command in addition to their own food flows. With new fossil-fuel energy subsidies the individual can do more work for his time, and a dollar buys more work. The price is lower. Local shortages of potential energy diminish the work done for the same time effort by individuals, thus increasing the price. The wages paid for an individual's work are thus related to the fraction of the work budget that the individual can obtain for use. An uneducated man in a modern system may have no means for controlling work processes with energy subsidy, and thus he cannot obtain payment proportionate to one who can do a job in which fossil-fuel power flow is involved.

The price of material goods is not determined by their energy content but by the work done by individuals in preparing them, plus expenditures they have incurred in obtaining services and materials. A study of the energy diagram in Figure 6–2(c) shows that the work services of a downstream specialist looped back upstream expend a small amount of energy compared to the total volume of energy transferred by the downstream. The money flows for both production work and specialized services are similar. Therefore, the ratio of money to energy increases downstream and is a measure of the increasing amplifier value of the energy there. Some work flows of a specialized nature may command more money because of the cost of accumulated information storage and training. Even more complicated are various other controls that cause individuals to price their own work differently. To some extent these various factors may cancel out for the system as a whole so that we can establish an over-all price for the system (Fig. 6–2).

The price system makes an adjustment so that the total amount of energy subsidy available for work is forced on the rest of the work processes equally. It serves to distribute the potential energy for work through the system evenly. When more energy is subsidized, the price is less and others are forced to adapt the same subsidy or be eliminated as a circuit.

Accumulations of money in a particular place result in faster spending (a special case of the flow in a circuit being proportional to the driving population of forces). Any slowdown in inputs may be balanced by an accumulation of money at the point of shortage. Most units provide mechanisms to loop more money back when there is a limiting factor (this is another case of energy loopback rewarding the input that serves to increase the downstream flow). Thus demand work is stimulated through the means of payment. Both effects serve as feedback. Adjusting the ratio of money spent to the work of the mechanism serves to return the system to a steady state without limitations at that point. Thus the price ratio of the payment operator serves as a self-stabilizing regulator. The price system is a decentralized local control mechanism which serves to eliminate local bottlenecks.

Demand is the work of pumping the flow in (see 3 in Fig. 6–5). The consumer is doing more of the work of the transaction and has to pay less.

A higher price stimulates the input unit to produce more; a lower price stimulates the output unit to draw more from the producer unit. Price thus operates like the feedback of work and minerals in ecological systems, stimulating producers when the ratio of loopback to forward flow is large, but stimulating the consumer when the ratio of loopback to forward loop is small. We may discuss price in ecological systems that lack money by measuring the ratio of the loopback to forward flow when both are in calories or one is in mineral terms.

WAGES, PROFITS, AND SAVINGS

Previously we have considered energy storages only as filters for taking the pulses out of a system or sometimes as instruments for setting up oscillating rhythms. Money storages are shown in Figure 6–8. Profit money stored as *savings* in a coffee can and not used for reinvestment has no appreciable energy value. If we burn the money without proof of its former existence, we merely change the ratio of money to work.

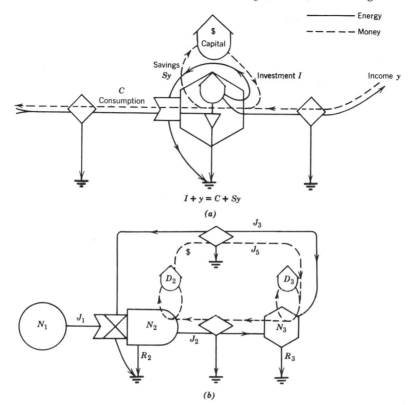

Figure 6–8 Energy-currency diagram which includes capital storages, income, investment, savings, consumption, and other features of the economic system. (*a*) Energy and currency pathways in a single self-maintaining industry, (*b*) pathways for a production and consumption economic system.

Burning the money merely shuts off a control circuit action that might have developed if the savings had been used through reinvestment.

Savings represent the money flows into storages for work services done and thus provide flexibility and time delays in the reward loop of energy expenditure. An equivalent in the ecosystem is the holding of fruits made in summer for planting in winter. The energy has already been expended in preparing a complex product, but an important material of low-energy value is fed back upstream and has the capability of developing new circuits. With the saving of both money and seeds, what is being used as loop reinforcement is critical information. In seeds it is in the genes; with money the critical information is already stored as the acceptance of the symbolic nature of currency as a measure of value by the network of people.

Wages are the money loopbacks in proportion to work done by the individual or by the work flow he controls as part of his job in an energy-subsidized system.

Profit is the money equivalent of the useful work done by a unit of a system in excess of necessary maintenance. A system unit has work flows from outside that do some of the unit's work, and money must be paid for these. Then within the unit there are fuel expenditures into additional work that require no help from outside units. These work outputs are the profit in energy terms. If the system is operating at a steady state, the output is sold with a price that provides money for the unit's payments plus some for the profit work. Thus profit has both an energy and money measure. When a system finds a new way to inject energies without paying for outside work, it reaps much larger profits.

The profit allows it to pay for more flows of work from outside to develop new circuits. Thus profit is the loop-reward method for creative selection of the network. Profits, if not creatively spent, do not induce more profits, and development of the unit then stops. Thus profit is proportional to the feedback of profit work into development. Growth can continue as long as there are new energy circuits to be tapped.

A government or other agency can multiply action if it arranges for work at an upstream point in the energy circuit symbolized by the multiplier block. Any increase of the quantity which was limiting there serves to allow proportionate increases of power flows downstream where the energy volume may be greater. This multiplier action is possible as long as the upstream work is opening a limiting factor gate over the range where the effect is linear (Fig. 6–7). It is not possible when there are no limiting factors other than the energy flow itself.

Capitalism is a network organization that provides for economically successful units to develop more structure in proportion to previous profit and thus prepare to generate more profit. The money received as profit measures the success of harnessing work to produce a system-needed flow. Profit serves as the immediate feedback reward stimulus for effective output. Capitalism thus allows its component units to develop their own networks for maximum energy flow and network complexity for maximum output. Ecological systems may maximize power flows into useful work in the same way. The more useful power flows provide more mechanisms of continuation against disordering influences, including those of competitive organization. Early systems of capitalism were like some disturbed natural systems with the separate units allowed to operate without central control. Competing units developed practices harmful for the system as a whole. Later systems of capitalism, like later successional systems in nature, achieved the same kind of maximization of the overall group function through the development of government and other interunit organizations. Thus in the more complex forms of capitalism and in complex ecological systems, loopback reward reinforcements operate on separate units and at the level of the whole system as well. The economic control is decentralized by having all the systems and parts receive their profits.

Some alternative systems such as communism provide for profits to be passed through a centralized governmental control system for reinvestment, thus in part eliminating the quick response time of loop controls. The automatic self-designing and self-regulating principle is thus discarded. The difference between the economic aspects of modern centrally controlled capitalism and communism is in the degree to which there is decentralization of initiative in feeding profits back into new development, design, and work important for the whole system.

Monopoly is the absence of alternative circuits. In a pioneer system temporary monopoly provides a focus of energies to pay for the extraspecial work requirements of new development. When huge power flows are necessary for the overall system, elimination of duplication can be economical. If a monopolistic circuit is limiting the larger network flow by artificially low output, the system provides for loop reward to set up alternates, thus regulating the falsely imposed limits.

A network can be manipulated in any desired way providing there are potential energies to spend for control work. Thus prices can be fixed, but moneys must be spent to prevent black market circuits. If survival of the overall system requires it, the economic system serves as a tool for supercontrols which stem from systems of greater priority to be discussed in the next chapter.

INFLATION AND DEPRESSION AND THE MONEY CIRCLE

The overall behavior of networks of nature and/or of man that incorporate the money cycle fluctuates, causing periods of inflation and depression. A principal aim of both ecology and economics is to develop enough understanding of systems to predict and manage them, especially the fluctuations, many of which have had disastrous consequences. We may discuss macroeconomics in relation to energy using the summary diagrams in Figure 6–9. Note the forward flow of potential energy into work and the circulating flow of money which is passing in a reverse direction through the payment operators controlled by it. The money circle serves as a kind of lubricant to complex branching energy flows and cross-linkages of work. Like ball bearings, the money turns in the opposite direction at the point of contact with the energy flows and work processes. Consider two states of the money-controlled system.

Depression is the slowdown in money flows caused by some failure in the money circle or other aspect of the system. With less circulation of money there is less money in relation to energy flow at each payment operator (the prices are low). Such low prices inhibit loopback stimulus

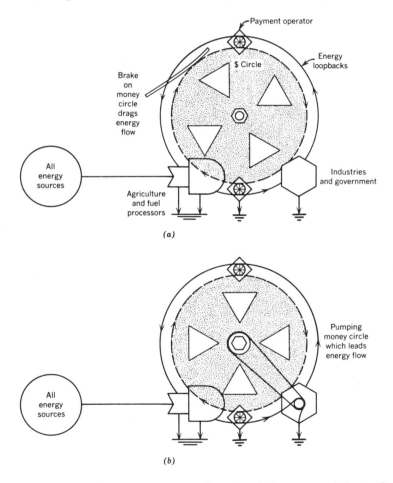

Figure 6–9 Energy diagrams comparing the role of the money circle to that of energy. (*a*) Depression with the money circle dragging energy flow; (*b*) inflation with the money circle being pumped to stimulate the energy flow.

of energy flows. With less energy flow the total amount of production decreases and the standard of living declines. The drag on the money circle serves as a drag on the whole system. Many kinds of factors apparently contribute to depression, including the disarrangement of society that follows wars.

An interesting posture of early American ethic affected attitudes toward the money circle and thus influenced its rate of rotation. Several generations of life in the simple economics of the frontier developed an appreciation for the value of work, of saving one's goods, of being economical with work resources, and of carefully planned diversion of some saved goods and work time into new farms, homes, roads, schools, and churches. The simple people had life-saving, realistic views of the energy circuit when little money was involved.

As systems became more complex and money began to play a larger role, the simple people made a fundamental error of transferring their realistic attitudes toward work, energies, and goods over to money, even though money is something different, mere symbolism traveling in the opposite direction. It was supposed, for example, that there was virtue in saving money in a tin can for later use as if money were potatoes or seeds. Behavior when bad times threatened was exactly backward from what was needed, thus serving to intensify depressions.

Inflation of money refers to an acceleration of the rate of money circulated in relation to the energy flow as in Figure 6–4(b). By circulating rapidly, money serves to stimulate energy flow, for the ratio of flow to energy (price) is high. Since work flows are greater, there is a higher standard of living. One process that stimulates fast circulation is the pumping of more energy into the kinds of governmental action that accelerate the money circle.

The fluctuations and cycles in business between inflation and deflation may be an oscillation produced by time lags in the relative response of the money circle and energy flows, each serving as a stimulus to the other. Adding money by printing it, or removing money by burning it, does not necessarily affect the relative speed of the money circle, although it does change the price of everything. Inflation or deflation by changing the quantity of money does not really affect the energy flow and the standard of living.

The work energies of the early Spanish colonial effort were directed at transporting more gold and silver back to Spain. This effort was not much different from printing more money and was vastly more expensive in work costs, draining much of the real energies of the colonies that might have been reinvested in development. The ethic of narrow greed did not stimulate proportionate loop reinforcement. A legacy of individual approach to problems still hampers the development of effective economies in some Latin countries, even though it is now common knowledge that complex group networks are required to process energies effectively toward a high standard of living.

COUPLING OF COUNTRIES IN WORLD DESIGN

Another important aim of energetic and economic studies is to find means for coupling two or more systems of separate countries to form harmonious combinations for the welfare of both. The world is undergoing rapid self-design as some of the advanced industrialized countries on fossil-fuel subsidy inadvertently disturb the underdeveloped countries still operating on solar energy alone. Some economically workable colonial networks were discarded in the fever of nationalism, without restoring equal replacements. What are the various possible connecting networks for coupling countries? Using the energy network diagrams in Figure 6–10, consider several kinds of intercountry connections of energy flow, work, and the money exchanges that may go with them. An industrialized country is shown on the left and an underdeveloped country on the right.

In Figure 6–10(a) the two systems are drawn economically separate with economic barriers. Both systems have their own closed loops, one that of a modern state and the other that of a primitive but workable state. The simple system left alone could continue as before. The underdeveloped country does not remain untouched, however, because with spread of information about systems elsewhere its people are dissatisfied. They may reach first for modern medicine which disturbs population controls. Forced by population growth they begin to experiment with parts of the production systems of advanced cultures and soon, with their original system disrupted, they must try for some other relationship.

If the rich country does nothing, leaving the underdeveloped country and its burgeoning populations to their own fates, confronted with starvation and desperation, some form of totalitarianism may evolve and set up its own crash programs, without the essence of democracy and peaceful development. Military ambitions may take the helm and provide even greater hazard to both the countries.

Another relationship between two countries is the tariff-protected economy symbolized in the diagram of Figure 6–10(b). A tariff is an energy barrier set up as part of a system design. The barrier in the form of various control gates requires energy expenditures, these being in proportion to the flow over that barrier and taken from the energy source that produced the goods being traded. Tariffs increase the energy costs of the system, but if they accomplish something essential for the network, their cost may be a justifiable energy drain of the overall system. For example, tariff protection may be needed if sources often fail. If the tariff accomplishes nothing for the system's effective future, it serves as an unnecessary drain that tends to diminish that system's overall chance for survival. If a primitive system has the energy resources eventually to support great diversity, the tariff may aid diversification and progress. Mexico may have this relationship to the United States.

A system, however, that carries considerable structure for contingency of stress may lose out if the stress occurs infrequently. Populations of organisms with specialization for deserts, arctic regions, or high salinity are eliminated in more ordinary conditions because too much of their energy goes to maintaining special networks of protection that are already provided from outside with no energy cost to them.

In Figure 6–10(c) is represented the simple combination of the economics of two countries. If, as drawn, these countries are at the same level of energy subsidy from fossil fuel and specialization, the economies may readily become one. Perhaps the United States and Canada are two such countries.

If, however, one is advanced in energy subsidy and industrialization and the other is underdeveloped, there may be serious difficulties. The rise in standard of living that goes with fuel injection into the food production economy of one country causes a lower standard of living among the subsistence farms and forces them to join the complex developing system. As we accelerate the money circle and raise the amount paid to individuals, the cost of something involving an hour's service work goes up so that everything to be obtained costs more. As long as one's work is part of the main system, the inflation affects payments received as well as money paid out. The subsistence farm that produces its own food and needs few things cannot buy these few things; prices of city goods have become inflated in relation to food because elsewhere food is being produced with fossil fuels. Unless there is a cash crop nothing can be sold to take advantage of the enriched main culture. The relative position is reduced. The cheaper food becomes somewhere in the world, the more the subsistence farmer loses.

The import of cheap food generates severe poverty for the nonindustrial producer of food and further reduces the food-bearing capacity of the underdeveloped country. With farms failing there is little incentive to make the investments necessary for injecting fossil fuels into the network to make cheap food. The more low-cost free food flows out from the fossil-

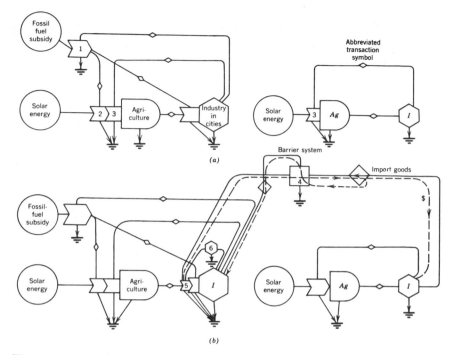

Figure 6–10 Various economic relationships between countries shown with energy circuits. (*a*) Two isolated workable economies, one an advanced fossil-fuel-subsidized economy, the other a simple agrarian economy. (*b*) Two economies with trade through tariff barriers. (*c*) Two economies of equal energy content linked with free trade. (*d*) Two economies with rich-country satellite relationships. The satellite acts as a specialist making cash crops and producing special manufactured articles for which low-cost labor is competitive. (*e*) A rich economy providing a flow of gift food to an underdeveloped economy and producing noisy energy dissipation by-products. (*f*) A rich economy exporting the equipment, knowledge, and starting materials of a complete workable system for tapping fossil-fuel subsidies into the underdeveloped country. (*g*) Two economies with different degrees of energy development organized into one system by a strong world government which is pumping total system yields into the weaker economy and serving as a regulator on all units.

Numbered work processes are as follows: (1) petroleum industry; (2) modern means of energy subsidization to agriculture; (3) loopback of the work of industry in cities toward high yields in agriculture; (4) tariff system including the work of monitoring and accounting—the flow of money is shown by the dashed line; (5) work process using output work of protected industry and imports over the tariff barrier; (6) protected industry; (7) energy of unemployed but fed populations dispersing in noisy, unconstructive activies; (8) work of research organization and export of complete workable systems with inherent loops for export; (9) world government control functions (see Chapter 7); (10) special programs for building up weak countries by using resources of wealthier countries; (11) world government staffs.

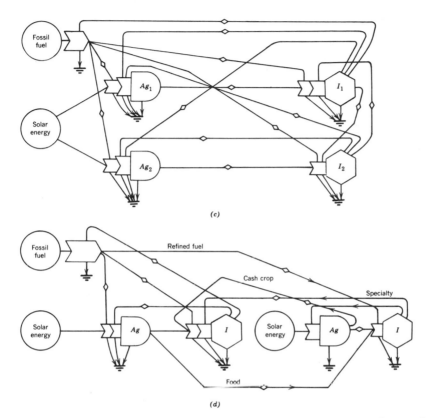

(c)

(d)

fuel-rich center, the more the underdeveloped country has to drop subsistence crops and substitute specialty cash products desired by the
productive system. What little wonder that economic iron curtains of all
kinds are put up to prevent the economies from becoming slave systems
and their food production capabilities from being permanently lost. Because the people in the towns can buy cheap food from abroad, they
desert their own market production systems and develop permanent
need for foreign food. If world food storages develop and prices go up,
with production in their own country having disappeared, there must be
a sharp drop in living standard of the towns whose progress was based
on fossil-fuel expenditures in the rich countries far away.

In Figure 6–10(d) the underdeveloped system produces a cash crop
because it can buy the foreign, industrially subsidized food cheaper than
it can make it, but to buy food it has to have money. Thus special crops

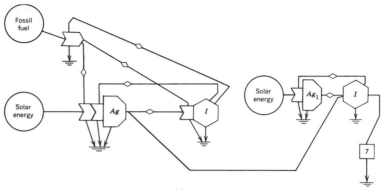

(e)

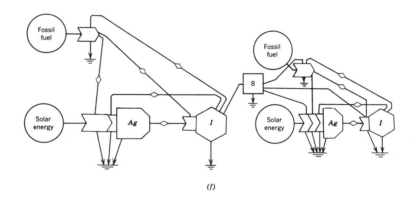

(f)

that have not yet been industrially subsidized are developed for export.
The two countries become engaged through the cash crop. With labor
costs low the underdeveloped country may be able to supply special
manufactured products too. The energy rich country, if it cannot or has
not learned to manufacture a special product, may be willing to pay ex-
orbitantly in relation to the food value or energetic cost of the product's
production and transportation. When this is so, it may be smarter economically for the underdeveloped country to produce the complex product
such as sugar, coffee, tea, chocolate, or spices, and use the funds to buy

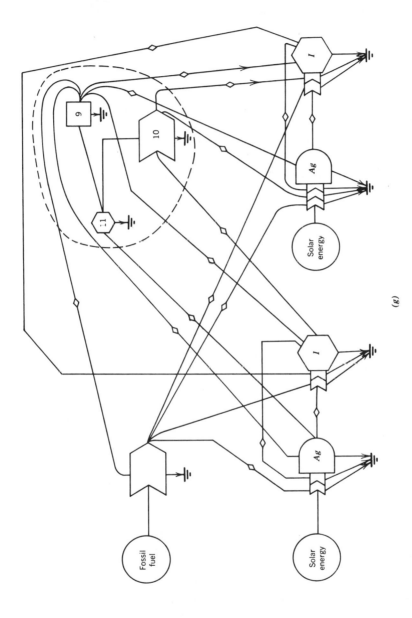

(g)

201

foods from the country that has the fossil-fuel already injected into its agriculture. This is the relationship of many tropical countries to the United States at the present time. They are economic satellites.

The hazard of such an economy based on scarcity is its collapse if too much production is arranged. If the tropical countries are disorganized and competing and if the specialized products are tree crops with long capital investments, market overproduction can have disastrous results. Such relationships are probably bad because they put the underdeveloped country in the role of a faraway specialist to a system beyond local understanding and governmental control. If food shortages develop the world over and cheap food exports from mass-producing countries disappear, the pressure to produce food will return. The standard of living will be lower unless by that time food production there is receiving some fossil-fuel injection. The sooner notice goes out about the limitations in food to come from now dominant mass-production lines, the sooner the change in plantings can be made.

Figure 6–10(d) shows the satellite relationships of an underdeveloped country performing specialty work as part of a rich economy. The poorer country must grow cash crops or engage in specialty manufacturing. The United States and Puerto Rico are an example of the circuit in Figure 6–10(d). Sugar and rum were cash crops followed later by manufacturing. This network shows the economic pattern of colonialism, but it occurs whether the countries are politically self-controlling or not. With common political control there is opportunity to feed downstream profits back into satellite development, whereas such feedbacks between countries without the same political organization require altruistic motives.

Figure 6–10(e) shows a relationship tried by the United States in 1966 in an effort to stimulate the economies of other countries, such as India, by giving food (or selling it cheaply). The fossil fuels and nuclear power are first passed through the advanced cultures for food manufacture, the food being exported as a gift; this pattern was discussed as a "short circuit" in Chapter 2. Such a system seems to leave control of the energies with the advanced cultures, but it is not a closed-loop system and does not receive the reward reinforcement necessary to form a permanent structure. Food sent out to the burgeoning populations without provisions for loopback reward has a negative effect on the producing system, and the receiving system develops larger and larger populations whose unchanneled energies either must go into random and hence destructive activity or, if focused into some group action, may have further negative effect on the parent or itself. To make such a food donation flow stable, the loop must be closed with a return of work, currency, and materials sufficient to make it stable over alternative adventures. If the loop is developed, a satellite economy results (Fig. 6–5(d)).

Another way to help an underdeveloped country is to design an entire closed-loop system with a fuel subsidy in the rich countries and export the complete loop process (see Fig. 6–10(f)). No food is sent, but the initial equipment and educational investments to start the loop within the underdeveloped country are exported along with an arrangement for supplying fossil fuel to that country. A closed, self-completing loop would include a food production complex plus one or more industries that produce insecticides, manufacture farm machinery, develop varieties, and process fuel. In the loop, food is supplied to the persons responsible for the work of the food-augmenting process. Such a development is not without hazard to the altruistic parent country since it involves setting up a fuel competitor, which in effect dilutes the relative power of the parent in the world. Aid for the development of Japan in 1935 was of this type.

Another alternative involves a single, fuel-based system of food production, closed loops, and a one-world economy (see Fig. 6–10(g)). No part is in competition if it is locked with direct and indirect loop-reinforcing flows and stable currency loops. This pattern may evolve from any of the previous possibilities. Since the larger systems tend to dominate the smaller ones, there is a tendency toward evolution of a one-world system. A world government solution may evolve but cannot really be initiated through the unilateral action of rich countries that control only limited parts of the earth's populations. The world system may require a strong central government or another form of coordinated government focusing reinvestment in the weak parts for the good of all. For the rich countries to give up their present power control to a system of untried capabilities may be very risky for the whole world's progress as well as for their relative positions. Some evolution in the direction of a network like that in Figure 6–8(g) is in progress. The United States government's economic relation to its states is a model of such possibilities.

MASTER SWITCHING CONTROLS

As we have seen in Figures 6–1 and 6–2, money flows are the feedback halves of loops that form systems in chain-like fashion. Money serves as an automatic regulating and equalizing device superseding the simpler mineral currencies and barter systems. Although money flows provide loopback reinforcement to control and integrate each energy flow, such money controls can be readily superseded by other control systems that serve as switches to focus and change the force network. In electronic systems many special kinds of relays, automatic response devices, and programmed hardware control and direct energies. In the ecological systems the behavior patterns of the large animals serve as power switches and transformers. For example, the flocking of birds concentrates power for migration.

In the human system the programming possibilities in people provide even fancier controls. Evolving with primitive man are many institutions stemming from his social psychology—his religion, his group nationalisms, his loyalties, and his loves. As partly inherited and partly learned pre-programs, these motivations are power-switching and power-concentrating mechanisms. They are means by which the resources and actions of vast numbers of people can become directed into one combined focus, resulting in the kind of power delivery that has been deterministic in the great wars, emigrations, social upheavals, and change. Thus Toynbee cites the role of religion as a major historical determinism.

These processes also must be consistent with both economic and energetic laws. The quantities of power flow developed depend on the available potentials, and the costs of the deterministic doctrines must have been borne by earlier expenditures of energy for their creation, maintenance, and inheritance. The power control system must in a flexible and adaptive way put power where it is required for group survival. Human behavior controls are limited to the energy capabilities of the systems sources.

To change system flows, taxes, loans, donations, and governmental spending programs can be manipulated using money as the means. The economic control system can also be bypassed with human means for controlling energy drives more directly. For example, in time of war a focus of power in an invading army may substitute a system of procurement and distribution not involving money. Being circular in its effects and readily controlled, money is rarely a primary determinant. More often it serves as an accounting method to measure the system's adjustments to energy-derived forces.

Because potential energy storage is the source of force and is a necessary part of all processes, including control actions, we may regard energetic determinism as the quantitative principle guiding events in systems. To the extent that the energy flows and derived forces are controlled by the circular system of money, we may also regard economic determinism as important. However, since master mechanisms serve to switch patterns into new energy networks as needed, we must regard local economics as the means and a constraint, but as just one of the determinants of network programming. In Chapters 7 and 8 we consider the energetics of the master controls of human systems—political institutions and religion.

REFERENCES

[1] Boulding, K., A Reconstruction of Economics, Science Editions, New York, 1962.
[2] Gambel, A. B., Energy R and D and National Progress, Office of Science and Technology, Office of the President, 1964.
[3] Ikusima, I., "Biology of Duckweeds with Special Reference to Their Growth," Physiol. Ecol., 10(2), 130–164 (1962).
[4] Sullivan, G. D., Profitable Levels of Rice Fertilization, Dept. of Agricultural Economics and Agribusiness, Louisiana State University Agricultural Station, DAE Research Report No. 340, 1965.
[5] Wilson, E. O., "The Ergonomics of Caste in the Social Insects," Amer. Natur., 102, 41–66 (1968).

7

POWER AND POLITICS

The true powers of individuals, groups, and political bodies lie in the useful potential energies that flow under their control. Power does work, gains and manipulates storages of energy, and directs forces. True power with a real energetic basis is a property that has no substitute. If some philosophy of government or some particular idealism attempts a form of control contrary to the hard facts of energy distribution, that system will fail. Real power and control must prevail in any test among systems, however stated. The energetic laws are as much first principles of political science as they are first principles of any other process on earth. Many of the political, military, and international problems of our times as well as the role of democracies can be clarified by phrasing them in power units. Energy diagrams for political institutions show energy controls and the constraints energetic principles place on their design. Some of these diagrams bear a close resemblance to networks of ecological systems.[1]

ENERGY CIRCUITS FOR POLITICAL ORGANIZATION

The organizations of man provide for decisions and actions that serve to open and close important circuits and control the rate of power flow in them. The acceptance by individuals of interindividual organization structure constitutes the essence of the circuits. Mental concepts form a

---

[1] Usual means for representing social and political power may not show clearly the connection of true energetic energy flow (power) and the manifestations of this at the level of human interactions (see, for example, Lasswell and Kaplan [13]). The essence of these theories do, however, seem measured by physical power. For example, deference value and influence are proportional to physical power flows gained.

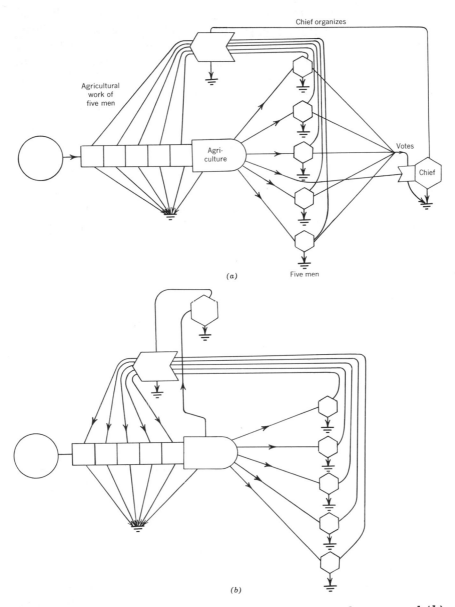

Figure 7-1 Comparison of energy network diagrams for (*a*) a democracy and (*b*) a dictatorship in a small agricultural system with a chief.

part of the structure. Through feedback pathways actions of individual voters on committees, of duly appointed authorities, in general elections, and through other institutions, control valves ensure continued flows of their own energy and those of the organizations they represent. Figure 7–1 shows some examples. The general form of the feedback is that of political institutions given by Easton [6] but in energy terms.

Successful institutions, like the other systems already discussed, have positive reinforcement loops which keep the control upstream working in a sensitive way to effect the best output downstream. A successful political organization is an energetic control that provides for energetic loop reinforcement. Successful systems of power budgeting require the positive programs of power delivery to necessary new activity as well as defensive budgeting to prevent power leakages to crime, useless activity, and unnecessary wars. Failure to control the system toward group-rewarding aims results ultimately in redirection of component power flows and reorganization of circuits to deliver greater loop reinforcement.

SIMPLE ORGANIZATION OF TWO INDIVIDUALS

An example of simple organization in energy circuit terms is given in Figure 7–2. Two individuals drawing on the same energy source are competitive. Then in (b), using their work energies, they test relationships. In (c) they adopt a pattern that gives one more control than the other. Energies of the conflict are released to go into feedback to control the members in a primitive sort of government. This simple organization eliminates competition and conserves energies and is able to evolve because the two units receive loop reinforcement from each other. The group performs as a better organized unit. As Berne [2] shows with a special language of two-way vectors, organization among individuals requires two-way stimulus-response, similar to the two-way flow of economic payment. As with the simple pecking order in chickens, there are also dominance relationships, preventing the separation of parts which, acting singly, might make conflicting and uncontrolled efforts. Dominance diagrams use vectors to indicate the direction of dominance. Dominance is the expenditure of extra work on control and is directed toward the work site. In the energy circuit the exchange operator is indicated as deriving its top-to-bottom work control from the dominant flow. In Figure 7–2 (c) flow 1 dominates and controls flow 2, but both are loop-stimulated by the other's work. The abstract pattern of Figure 7–2(c) applies to many relationships, such as love, partnership, contract, and symbiosis.

SOCIAL STRUCTURE AND POWER

Systems of higher animals and man through their programs of information transfer and response to each other's information exert controls and
influences on each other's works. These are adaptive and serve to produce
group actions corresponding to group structures that develop. In psycho-
logical studies the principles of social action through power and force
have long been recognized and described with factors and matrices of
different combinations of interaction (see, for example, Ref. 4). Recognized are resistances to forces, opposing forces, and some of the same
concepts which have their counterparts in the physical and biological
systems. For historical reasons social science has been reluctant to recognize that the pathways and expressions of social power are the same
kind of flows that occur in electric power lines and atom bombs. The
measures of correlation, frequency, and probabilities have been used
instead. The energy diagrams may encourage the use of a real energy
unit (kcal) as a common denominator, uniting physical, biological, and
social theorems.

In Figure 7–2(d) a human individual is shown with opinions entering
and leaving and with actions resulting from the opinions prevailing in
the information storage. A social structure of three such individuals
(such as executives in a corporation; members of a family, etc.) is given
in Figure 7–2(d) forming a power network like that discussed by French
[7]. In this example, the energy flows include those of the human operating on food supplies plus larger power flows that depend on their actions.
Included is one pathway of data from the real world (not opinion).
Truth is the state of noncontradiction, in this case between the several
inputs from real world and opinions. As shown, each opinion pathway
is really a double pathway including work of transmitting the information and the tiny energy content of the information itself (which has high
amplification value). The degree of influence of one unit on another
through opinions includes the interior matching of truth plus the coefficients of the input multiplier actions that represent the previously established status of the other individual.

DIAGRAM OF A MORE COMPLEX TRIBAL ORGANIZATION

In Figure 7–1 a small tribal organization works on agricultural plots
where their group action requires coordination for maximum output of
food. The food flow supports the individuals who choose a chief through

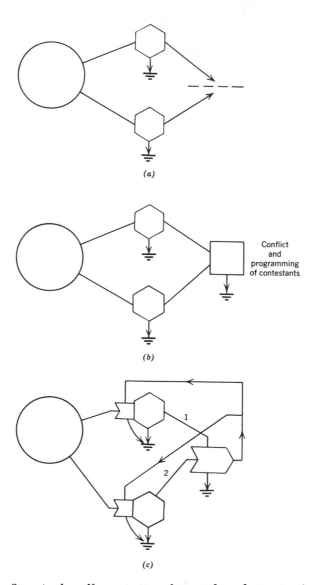

(a)

(b)

Conflict
and
programming
of contestants

(c)

Figure 7–2 Steps in the self-organization of two independent units. (*a*) Two units drawing on the same energy source do work to locate contact. (*b*) Two units do work in establishing a relationship, for example, conflict. (*c*) Two units are linked with one dominating the second, contributing joint reserves towards regulation of both. (*d*) Detail of a human's role in receiving opinions into an information processing and storage (*I*), transmission of opinion, and actions of work based on opinion in storage is dominant in the control mechanism set up in (*c*). (*e*) Social power system example from Harary in Cartwright [4] drawn as an energy network.

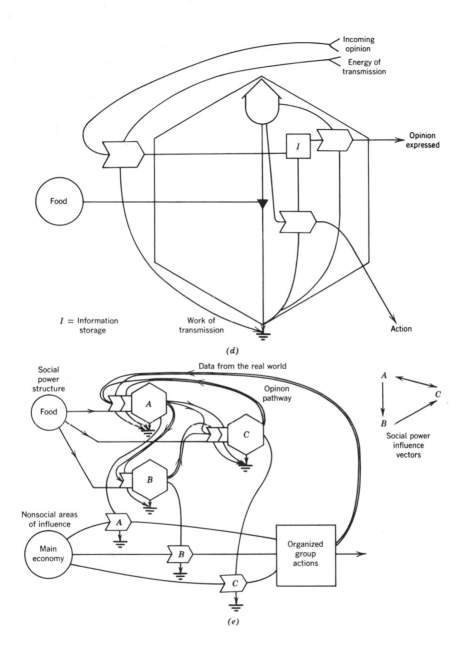

Incoming
opinion

Energy of
transmission

Food

I

Opinion
expressed

I = Information
storage

Work of
transmission

Action

(d)

Social
power
structure

Data from the real world

Opinon
pathway

Food

A

C

B

Nonsocial areas
of influence

A

B

Main
economy

Organized
group
actions

C

A

C

B

Social power
influence
vectors

(e)

a small work expenditure for organization of the group, the exchange of opinions that sets up acceptance of control in their memory systems, and finally the selection of the leader. The chief then does the work of guiding the group. As shown in the diagram, such a system has loopback reinforcement work for the individuals in relation to agricultural production, and the chief helps to prevent competition and to focus group effort—a necessary system expense. In the process of system self-design some extra work may have to be expended in fighting, for dominance is measured in proportion to the ability to control others through physical power in some primitive circumstances.

In better circumstances when individual physical power is no longer the measure of the ability to control others, voting procedures are relied on and provide a more accurate accounting of the number of component powers available. Control mechanisms must recognize the true distribution of power or be by-passed. They will be loop-rewarded only by drawing most of the power of the system into group actions. Successful stable political organization must represent actual power distribution or be eliminated as some of the excess power in alternative patterns flows into the work of reorganization.

DECENTRALIZED POSITIVE LOOPS OF DEMOCRACY

The organizational pattern of the tribe in Figure 7–1(a) illustrates some important characteristics of democracy as a successful network design. Compare the differences in design of parts (a) and (b). The vote control over the chief provides a decentralized group of controls which record and express the degree of loop reward developing from the actions of the chief. Thus power does focus for action in the chief, but it decentralizes further around the loop for verification and control of the control. The voters through the voter-control gate serve as a low-energy means for enforcing power deliveries to themselves.

In Figure 7–2(b) there is no voter control and a dictatorship pattern results. [The flows may be controlled away from the welfare of the people from whom the power is cascaded.] Reorganization takes place with the component powers redirected. Thus the simple dictatorship is not a stable system if it cannot obtain some kind of loop control to sense when the actions upstream are producing the best downstream consequences in useful energy deliveries.

The simple democracy in Figure 7–2(a) resembles an ecological network. In our discussion of natural systems in Chapter 3, we found that the complex higher organisms in their own metabolism processed a moderately small part of the power budget of the whole system, but that they served as controls on the less complex power circuits. In the natural system power control is not, however, focused in one entity. Some locally overpowering units such as the tiger or the killer whale are capable of immense concentrations of effort to accomplish some large-scale activities such as regenerating minerals from the larger animals, but their power is still a small part of the total power budget of the system since such species are necessarily widely distributed. The most stable of the natural systems have great diversity in many complex species and thus much decentralization of control.

According to modern principles of successful management, the industrial system also has its energy control decentralized. Top management has the power concentration to accomplish the large-scale work functions a few at a time, but in terms of total power being controlled, the main volume of decisions is decentralized throughout the industry in various relays and through human foremen.

The age-old argument about which society is best, the highly centralized one or the decentralized one, is resolved by the general principle that there must be some power concentration to accomplish whatever large-scale acts are necessary, but the bulk of the decisions and controls on small-scale processes must be decentralized among the various smaller units of government. A heavy reliance on individual decisions both large and small is necessary. We call such power control distribution at many levels democracy.

POWER BASIS OF PARKINSON'S LAW

Almost everyone has heard of some of the ideas about organization that carry Parkinson's name, all to the general effect that efficiency diminishes as the number of units in the organization increases. Sometimes these statements have implied that such loss of individual efficiencies is unfortunate. If, however, this loss is the natural cost of organizing for the accomplishment of work that cannot be done by individual parts, we may accept a small power diversion as a necessary cost of order, as a power tax in other words.

The development of circuits and controls that connect parts into a whole network is called organization (see information concepts in Chap. 5). One individual, species, or group is coordinated with one or more others by functions that involve a network of materials, energy, and work process. Such organization processes take energies that might have gone to the individuals. The animals in a forest or a sea use energies in their behavioral interactions and, as populations, perform certain activities in unison. Fishes school, birds divide up a food area by zoning themselves in territories, and insects divide up the feeding times with some operating at night and some during the day. In organized society 20 to 50 percent of each citizen's dollar budget goes into supporting the government.

In low-energy systems there is little energy for organizational governmental activities because there is hardly enough for individuals. In forests the larger hawks and the migrating flocks of animals and birds compose a weak government, integrating the energy flows, cropping excesses wherever they tend to develop, and redistributing the minerals involved in their waste dispositions. We may speculate that energetic paucity prevented primitive activities of man from developing a high order; temporary accumulations of power permitted some expansion of organization but it always disappeared when the temporary potential energy became unavailable. The present tight organization of large areas of the world was not possible before fossil fuels became a major subsidy.

As long as energy budgets came from the sun and were small, the ordering of larger groups was possible only by reducing to a minimum the energies expended by the individual for his own use, as in the slave systems of Rome. A new discovery such as a military innovation often gave a group temporary superiority in its ability to control the power flows of others, but then use of the innovation spread.

In the past centralization of power and individual freedom in using energy were not always compatible on a low budget. Man was able to organize large areas of the world only when large excess calorie inputs became available for integrating systems of communication, governmental organization, and military outlays as well as for prerogatives at the individual level.

To survive and maintain a competitive position, a system must draw the maximum power budget possible in a situation and process this budget in works that reinforce future survival and stability. Surviving systems apparently have much decentralized control but some concentrated power capabilities. As the number of compartments (populations, individuals, social groups, etc.) is increased, the number of possible pathways rises very rapidly according to the graph in Figure 7–3. (See Figure 1–7 for the possible pathways to provide one between each unit.) Also shown is a graph of the number of pathways needed to provide forward and return pathways between each. Complete organization might be defined as the arrangement of energy circuit relationships among all

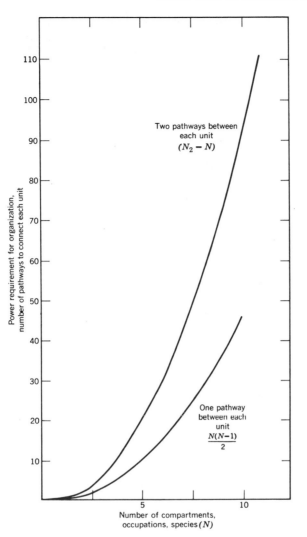

Figure 7–3 Graph of the number of pathways needed to connect each unit with every other unit, with two pathways and one pathway as a function of the number of units (N). Power requirement for organization may be proportional to the number of pathways required.

population units. Most situations are not this fully organized. The network example presented in matrix form as Figure 1–7 showed the difference between the possible pathways of relationship and those actually operating. The number of possible pathways is a measure of the uncertainty and opportunity for confusion of functions. The number of actual pathways is a measure of the organization and complexity of function. The amount of organization that is required for different situations is a good research question. If the cost of adding new circuit junctions goes up in proportion to the number of connections, we may plot the amount of complete organization that is possible as the energy for organization is increased. The larger a system's energy support, the more energy it has for organization. Thus the potentialities for diversity are related to the energy budget in functions like that shown in Figure 7–3. E. P. Odum [14] with Menhenik found species curves following square functions. Figure 7–3 suggests why this might be so.

Many low-energy systems have achieved a stable existence for their people, and many low-energy systems may resist the influx of the ways of the power-rich. Simple cultural characteristics may be kept temporarily. Perhaps unfortunately, a low-energy budget system is ultimately displaced by a system that can involve more power and more work processes, as long as the energy sources are available. The energy-rich system can by control decision preserve the lesser one at its own expense if its own power needs are already covered. This has often been attempted by modern man with arrangement of native reservations.

FREEDOM, INDIVIDUAL UNIQUENESS, AND POWER BUDGETS

In history the long effort to develop individual freedoms was part of the struggle for control of the system's power budgets. At first power control was completely decentralized because little or no energy was put into the organization. Consequently, the individual was completely at the mercy of fluctuations, environmental factors, and the many phenomena that had substantially larger power budgets. Complete individual freedom to control one's fraction of the power budget was accompanied by poor opportunities for survival and enslavement to environmental dictatorships such as poverty, famine, and disease. Groups of unorganized individuals became incorporated into other systems, some of whose power budgets were applied to larger-scale competitive group functions. Succession in natural communities proceeds from initially isolated individuals toward a network organization. Separate individuals cannot specialize, and without organization, functions are inefficient; neither capital nor knowledge can be accumulated, and no progress is made.

At the other extreme, when very little of the group budget is available to the individual, great centralization of power may result in large-scale operations, but there are then few mechanisms for performing the small tasks effectively or for making use of the decentralized capacities of the individual's mind. Unfortunately, the historical fight to prevent overcentralization may still carry the cause of freedom to the extreme that dictatorship of the human system is reverted back to dictatorship of the environmental whimsy.

When systems allow too many individual power prerogatives and not enough of the individual's power budget is directed into organization or coordination, each individual, being unspecialized, is almost one of a mob. In that state individuals are interchangeable units of a population. The individual may have freedom from organization, but he has no special value, no special mission, and he forms no special and necessary part of the energy flows. As an individual he has little ability to tap the fossil-fuel flows. The raw consequences of too much freedom are little respect for the individual and few chances for a stabilized future; all the values normally implied in progress are absent. In natural environments a system of units, all of the same species, many constitute one power circuit with all individuals alike, regimented in population and operating without ground controls. Such communities, however, are without future except to be replaced by more complex systems. They may be good for initial colonization and hence for one round of primitive flourish before displacement by a higher order.

Thus basic energetics require some loss of freedom of control over one's power budget in order to gain freedom over the pestilences of the environment and to receive injections from the rich flows of fossil fuel which cannot be utilized by simple individual systems. If energy is added with each individual so that the power budget goes up with the number of individuals and if some fraction of this budget is taxed for organization and government (such as 25%), the number of organizational categories and divisions (compartments) that can be added increases much more slowly as shown in Figure 7–3 and increases more slowly than power for organization.

For loss of freedom the individual gains something vastly more important to him—the opportunity to specialize and become sufficiently different and singular that he can make a unique contribution. Specialization in advanced natural systems in modern society so dissects the networks that there may be only a few of a species or of a rare occupation. The opportunity is great for an individual to make an important contribution that would not be effected if he were not there. There are so many possible pathways that only part are used. The stronger the organization, the more opportunity there is for a unique individuality within the specialty. However, the number of jobs is small and freedom to move is thus restricted. The value of the individual is in inverse relation to his freedom.

So little understood are such concepts that the uneducated mobs of the underdeveloped areas and subcultures seek the state of high freedom from organization that carries them diametrically away from what they really want. In the disintegration of some former systems an anarchistic state of freedom is attained, like the first stages of some ecological succession, but it is rapidly taken over and organized because the stability and lasting power of the whole group is enhanced when group power coordinates the individuals with feedback loop controls like that in Figure 7–1.

The urgent issue in the power struggles of the 1970s is who and what will be the organizing systems of the anarchistic areas. Those who work to give areas freedom or urge uneducated peoples to seek so much individual freedom wrongly carry these areas back into starting succession again. Nations are made great by their organization, and it is this order that can be exported and taught, allowing the increased power flows, the better specializations, and the rising value of the individual.

Once a system design has been evolved, the energy cost of evolving it again is large compared to transplantation. Some democracies have liberated their colonies without an organization, leaving them with the alternatives of slow evolution or transplantation of some other preformed organization.

GOVERNMENT OF SATELLITE GROUPS

An example of a political organization that develops tensions because of the failure of loop reinforcement to develop local services is the colonial government pattern (Fig. 7–4). The parent system has a satellite at a distance over which it exerts control, but only in proportion to the service rendered back to it by the colony. The parent country appoints a regent to head the colony's government and controls the energy contributions (economic control) it makes to the colony.

The individuals in the satellite produce in a loop that best rewards the parent, but there is no mechanism for loop reward of these services to local welfare. Thus tensions develop between competing control loop pathways, one local and one to the parent. For example, university students want to run the colony while the parent population, acting through its regents, guides it in other directions. Sit-ins are the consequence. Boston tea parties resulted in the American Revolution.

The ultimate solution is in the distribution of power. When much of

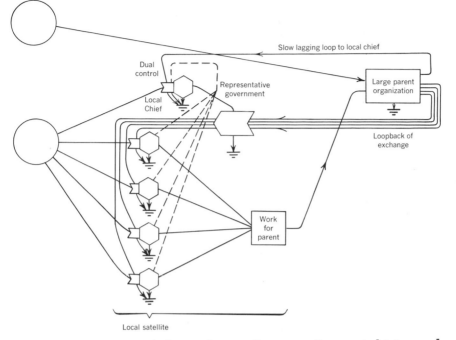

Figure 7–4 Energy network diagram for a satellite system. Loop control is improved by adding the representative flow (dashed lines) partly controlling the chief.

the power within the colony is generated from agriculture or fuel-based industry, the satellite may separate if it has enough excess energy to cut off the control functions of the parent. Thus were many colonies made independent, either through revolution or threat of it. However, if most of the energy is derived through the parent, the satellite group cannot separate. Tensions can be relieved and partial loop reinforcement provided, however, if a loopback route for participation in government is added, as with the dashed line in Figure 7–4. Thus some representatives are included in the decision control work. The principle here is to make representation in proportion to true power control.

When work is done by one process on another with work gate operations, large amplification factors are important if the first is to control the second. When the work operation is one of disordering the other system, it is operating in the direction of the natural tendency for disordering and dispersion of the second energy law. Less potential energy is required to disorder than to order. An illustration of this is the role of light energy in plants in two processes, photosynthesis in which light builds up structure and photorespiration in which light is harnessed to help the disordering of respiration. Kowallik [10] found much less energy required for disordering than for reordering. It takes more police to control criminals. Whenever disordering is adaptive, it can be done cheaply.

In war disordering serves to prevent the action of the warring system and much damage is accomplished for very little energy. Perhaps the prevalence of war in some earlier systems for such purposes as territorial organization and renewal was actually a form of economy. Now however, with vast new quantities of energy injected into war programs, the system has shifted to one of excess waste. A war zone develops a high energy flow since it must receive the energies from both systems both for competition and for control. For example a forest edge is an ecological boundary that has this property. It has more birds. Some complex ecological systems which involve great adaptability and resilience have ecosystem soldiers that operate at the ecological system level of organization. As here used the term soldiers refers to specialists whose biological behavior programming is focused on destroying those outside agents which are of special danger to the parent ecosystem. The kelp beds of the sea raise kelp bass which are programmed to concentrate on urchins that would otherwise cut the kelp and float it away. The fishes of tropical reefs will graze back turtle grass in the vicinity helping to prevent the competition for the mineral cycle of an alternative system.

POWER MECHANISMS OF CHANGING NETWORKS, VOTING, AND WAR

Man's ability to incorporate new power flows and to concentrate them through organization into military power has grown along with his other structuring skills. In early times when man was energetically a small part of his supporting ecological systems, his wars served as part of the means for developing organization beyond that of the local group. When tribes engaged in indecisive conflicts, the wars served to maintain territories, achieving a division of the resources, which is a form of organization into groups. When these wars were decisive so that one group came under the control of another, an organizational structure of greater dimension was imposed. War at this stage constituted an energy flow into organization.

Later other means for organization with new institutions and modern communication methods were developed. These organizational means transmit energy flows into forces through internal controls such as police and costly education. There is an interesting fundamental question: which is energetically cheaper, the energy budget for a world that self-organizes through war-established boundaries and occasional territorial rearrangements or the energy budget of a fully organized world that keeps the same degree of order through its internal police?

When the energy flow in human-controlled pathways was small, there was only one kind of circuit. Tests by these circuits of each other's ability to control are part of the process of setting up a network as in Figures 7–1, 7–2, and 7–4. Now, however, there are enormous power circuits of whole countries and large groups as well as the small power circuits of individuals who represent and govern the large groups. What has carried over to modern conditions is the practice of determining control hierarchy by testing power. When the power involved was small, the cost of organizing the group was small. Now that there are huge energy subsidies, actual power tests of control are immensely expensive.

Figure 7–5 shows two stages of testing the hierarchy of control, one with a complete group and one with representatives in the hierarchical efforts of political bodies. If, however, these proxy tests are to be realistic and truly represent the group's abilities in real power tests, the voting must be in proportion to the true power of the group. Having representative people interact so as to predict the behavior of larger groups they represent is a well-established experimental procedure called social simulation.

AN ENERGY BASIS FOR THE UNITED NATIONS ORGANIZATION

Our aforementioned premise is that voting power must correspend with real power. Organizations such as the League of Nations whose real power did not correspond to voting power were soon bypassed. In 1966 the equal voting strength possessed by countries of small energy budgets and those of vast energy budgets caused some world disputes to move outside the United Nations, degrading the organization's true might to that of a paper tiger. A world organization can be restructured by a simple procedure which is consistent with energetic principles and which might bring about world peace.

The power budget of a country is readily computed and could be determined annually. Instead of the yearly assignment of votes in proportion to population, they could be assigned in proportion to the energy budget of the previous year. Thus the voting power would always correspond to real ability to influence the world. Since economic power and military power stem from power budget, these influences would then

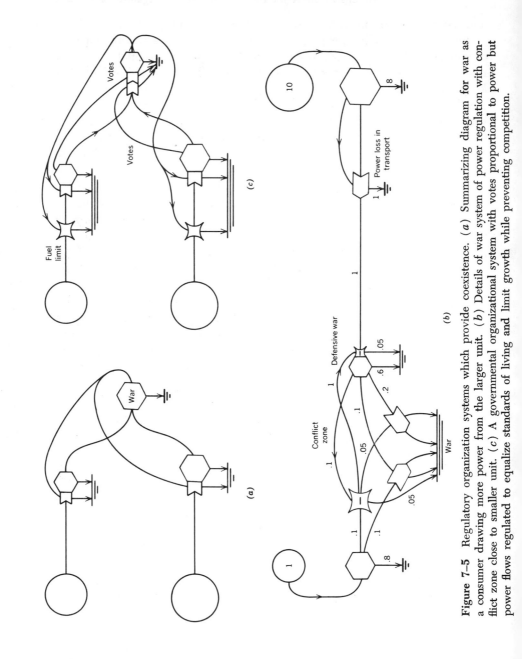

Figure 7-5 Regulatory organization systems which provide coexistence. (*a*) Summarizing diagram for war as a consumer drawing more power from the larger unit. (*b*) Details of war system of power regulation with conflict zone close to smaller unit. (*c*) A governmental organizational system with votes proportional to power but power flows regulated to equalize standards of living and limit growth while preventing competition.

correspond to the vote in the world organization, thus allowing a voting demonstration of power without the vast expense to the general welfare of a direct economic or military test. By this means any majority vote is automatically backed by a power excess, and there is little temptation to ignore the majority will. Furthermore, any attempt to ignore the majority can be prevented through the direction of its power into force channels, since the power reserves are available.

The power budget means of allocating voting power has flexibility so that an underdeveloped country which increases its power budget with industrialization gains votes in proportion. The power budget includes all kinds of energy. An agricultural nation with few fossil-fuel subsidies has most of its energy in its metabolic budget, in the calorie flows of the individuals and the farm animals. To this we must add the consumption of industrial fuels starting with those derived from the solar-energy-based systems such as wood, and including the fossil fuels and nuclear power. In Table 7–1 is shown a computation of votes in proportion to energy budget. Very tiny countries that are not industrialized will not have a whole vote and will have to share representation with other countries, whereas small but powerful countries can vote in proportion to their impact on world economy and military budgets.

The process of annually computing votes from power budgets for everyone to see also serves to eliminate from power politics some of the attending apprehensions, propaganda, and sabre rattling, for true power is known as it develops without military and economic demonstrations. The world is going to run according to true power anyway, but the voting arrangement allows it to be done without wasteful wars and economic bludgeons.

To stabilize the competition of expanding economies for fossil fuels, limit growth, and still provide justice to members of less developed systems requires additional features of power regulating system. Power limitations by the central government must be constitutionally allocated on criteria that will equalize the standard of living without providing power for growth and expansion. Some such system could stabilize population levels, bring populations into similar energy levels per capita and do all this with the energies formerly paid to the conflict as an organizational mechanism. These ideas are presented by the network in Figure 7–5(c).

Ardrey [1] recounts the behavior programs of animals and man distributing themselves over the land. He correctly indicates behavior as the mechanism for an adaptive purpose rather than the ultimate cause. Stating it in other words, human behavior makes sense only at the system

Table 7–1   Ratios of High-Quality Power for Allocating Voting Rights and Tax

Country	Area (10^{12} m^2)	Population (million)[a]	Solar Energy Contribution of High-quality Energy to Man (10^{12}kcal/day)[b]	Fossil-Fuel Consumption (10^{12}kcal/day)[c]	Total High-Quality Power (10^{12}kcal/day)[e]	Assigned Votes and Tax Quota[d]
United States	9.52	201	37	33	70	83
U. S. S. R.	22.3	239	61	11	72	85
Japan	0.36	101	1	2	3	4
India	3.18	523	15	0.4	15	17
Brazil	8.51	88	35	0.5	36	43
World's land	149	3479	508	58[e]	566	670
Sea authority[f]	361	none	280[g]	—	280	330[h]
World[i]	510	3479	788	58[e]	846	1,000

[a] 1968. Population Reference Bureau, Washington, D.C.

[b] The solar energy contribution to man as life support and directly as organic matter of fuels, foods, and services is that not being subsidized and augmented by fossil fuels. In absence of detailed data this column was very approximately computed as 10% of the photosynthetic production. For the land photosynthesis was taken as 1% of sunlight of the regions [15]. See Table 3–3 for representative efficiencies and Sellers [15] for regional sunlight.

[c] Calculated from per capita data for 1961 in Fig. 6–6 [3].

[d] Based on the world high-grade energy flows as 1000.

[e] See Table 2–2.

[f] An authority representing the seas may have a vote proportional to services from the sea and fossil fuels consumed there.

[g] Direct yield of 50 million metric tons of fish [5] is small (0.1×10^{12}kcal/day), but the ocean provides major services of high-grade organic processing as a life-support system. Ten percent of oceanic production from Hutchinson [9] was taken in absence of detailed data. See note about oceanic productivity estimates in Table 2–2.

[h] Votes assigned to a sea authority on global issues or those concerned with the sea. Life-support service serves in lieu of tax.

[i] Land plus sea.

level, individual responses being part of programs of overall adaptation. Formerly "the territorial imperative" as Ardrey calls it served to distribute the populations of man evenly in relation to the energy incoming from the sun which was more or less area proportional to area. Now, however, the extra new energies of fossil fuel and nuclear power are not coming in on an area basis but are available to any group which can hire a tanker. As Table 7–1 shows, the solar energies are still important but are now augmented by fuels. Our human inherited and cultural behavior may still be oriented to the old need and contribute to the poor adaptation of man to his planet. Individual and institutionalized behaviors need to be oriented to distribute all power inputs, not those of the land only.

UNLOOPED ENERGY CIRCUITS AND SOCIAL EDDIES

Like murder, so the expression goes, energy will out. A flow of energy either goes into storage or drives processes. If it goes into storage, it builds up its ability to drive more processes in that place. If a flow of energy does not do useful work and does not have loop controls, that energy flow may do work that is detrimental and explosive. Laws of energetics limit not only the nature of useful work but also our ability to discard potential energy quietly.

In electrical systems short circuits are a flow of energy not harnessed to useful work. Instead, the rapid dispersion of energy into heat forms heat gradients and does detrimental work such as melting the insulation of wires. In forest fires sudden release of heat develops strong thermal winds, a pattern which is equivalent to river systems in which the waterfall throws its potential energies into geological work on the riverbed. In the atmosphere sudden absorption of energy from the sun produces heat gradients that set up convection clouds which in turn maintain eddies in the air. In all these examples the proffering of potential energy leads to dispersive processes that also do work as they shunt away energy. Energy cannot be released without doing work. Hence any design or energy networks must provide for a controlled and useful channel for the energies supplied or the system will find its own circuit, one possibly inimicable to man's preferred pattern.

In social systems we observe the proffering of energy flows in ill designed mechanisms for welfare support of idle youth groups, the unemployed, and in other arrangements that sometimes prevent potential energy from flowing into useful work and may not provide reinforcing loopback control. Then, as in the short circuit or the forest fire, the energy emerges in useless eddies, in this case of a social nature. Populations supported by labors other than their own may direct energies in random directions and into such disorganized activities as mob agitation, disorganized play, theft, military adventure, and disruption of previously existing economic loops. These activities spend the metabolic energies, but they may not do the useful work on the system necessary to form a self-reinforcing loop and stability.

If excess energy flows continuously, a natural selection of activities that are reinforced into loops might be expected, but if the energy flow is intermittent or irregular, such organization is precluded. In natural pine forests a fire is part of the stable system as long as it is regular and helps release minerals and restart the succession of tender plants. Social eddies can become part of the stable system of things provided they are regular and serve some feedback-reinforcing functions. Simple war and energy-using games have had that function in past systems. If means for the regular use of excess energy storages are not provided, they may accumulate until the fire, explosion, or social disorder is so great that it is destructive to the whole system. The elimination of fire from pine woods sets the stage for disastrous fires twenty years later. The elimination of energy-shunting games and drains from energy-rich economies may have the same effect. The apparently frivolous pursuits of sports and hobbies may have important roles in the democratic systems, preventing misdirection of energies while providing a divertible storage.

LAW

Previous discussions have shown the importance of holding a workable network. As systems become old and complex, special energies may go into conserving network features through the institution of law. Law may be defined as the formal statements of a systems network and its switching alternatives. It is the group's means of maintaining its network in a workable design.

Using Hohfeld's theory of dynamic law, Hoebel [8] has examined primitive cultures for the emerging patterns of law. The essence of law as described in these simple systems can be identified with energy network concepts. Consider the four types of interperson relationships of the Hohfeldian system and the various parts of Figure 7–6. The italics refer to the terms used by Hoebel for the eight basic relations. In Figure 7–6(a) the "demand right" of one person to draw a duty from another may be identified with an energy circuit upon which the downstream recipient may at its own discretion elicit some upstream work or pump flow. The upstream unit may not break the flow.

The "privilege right" of one person to operate without a demand right from the other is an energy flow without connection from the other. It is an insulated relationship which only the privileged may change with switching action.

In Figure 7–6(b) the "power" individual may generate a work flow involving a liable individual with or without a contract for reverse loop-

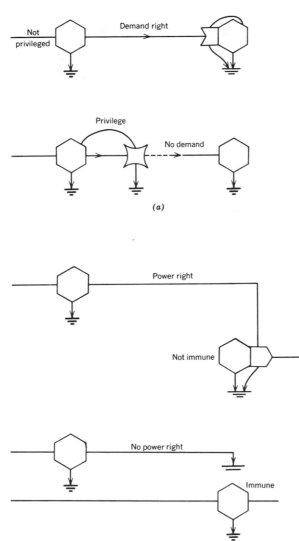

Figure 7–6 Energy connections between two parties which correspond to legal rights. (*a*) Demand right or privilege; (*b*) power right or immunity.

back. The liable individual may form a flow only with the switching action of the power individual.

The "*immunity*" of a unit to any *power* flow from another unit affecting it in network terms means that the immune unit has no network connection and has the right to remain so insulated.

LAND POLICY AND POWER GENERATION

One of the worst instances of political trends running contrary to basic energetics is the fever of some land reform in underdeveloped countries. As we saw in Chapter 4, the basis for really large increases in agricultural production has been the coupling of the system to an industrial economy so that heavy fossil-fuel power budgets are directed into food production. Such injection requires fairly large systems and considerable initial power diversions into equipment, organization, and know-how, usually described as capital investment. Considerable blocks of land must be involved to make the industrialized agriculture based on fossil fuels effective. What may be needed in Latin America and Africa is more organization of production rather than fragmentation into individual unrelated units attempting to operate on a solar budget. Fossil fuels can subsidize small farms, for example; tiny tractors are used on small plots in Japan, but the small farm must be part of an organized industrial economy to take advantage of fossil-fuel subsidy.

The fever of land fragmentation results from a confusion between ideals for production and power generation and ideals for power distribution and control. There can be no question that the system generating the most useful power for useful work is the desirable one and the one that will ultimately predominate. The problem is to set up the best system of power generation without the long, wasteful trial-and-error selection, and to discard processes that had to be gone through the first time there was innovation but should not now have to be repeated in each new place. The land policy for developing maximum useful power has little to do with man's need for land to live on and his pride in his dwelling.

Many lands have storages of minerals and soil structure and can be immediately developed for intensified agriculture. The ecosystem and climate have made capital investments in the land in the form of storages and other work. Other soils and lands, especially in the moist tropics, lose what storages they have in a year or two of agriculture. The natural systems there store their mineral requirements in their leaves and dead ground litter and branches, recirculating the minerals as fast as the microbes release them. Thus clearing the land destroys much of the natural capital. For agricultural use we must supply the minerals and soil structure or develop a system that retains the natural mineral cycle. The tree umbrella type of farming, consisting of a crown tree and a substory tree with one or both bearing a product useful to man, is common in the moist tropics. The crown tree continues to cycle the minerals, hold the soil structure, and shade out the weeds, but some power is diverted for this service and less energy goes through to the ground tree which yields products such as cocoa and coffee. Other agricultural products from tropical tree systems are Brazil nuts, rubber, tapioca, tea, and fruits such as mango, breadfruit, and banana. Tree crops cannot be developed in a single year; on the contrary, years of capital investment are required. Because larger organizational energies are needed, these areas are difficult to develop in small plots if there is not a ready connection to a large power subsidy.

THEORIES OF HISTORY

Any citizen conscious of the rapid sweep of modern history thinks some-times of the rise and fall of civilizations and their causes. A civilization, like a forest, develops organization, specializations, and the structural manifestations of concentrated work. Some theories of history also involve a cycle of rise and decline.

N. Danilevsky and O. Spengler tried to find youth and senescence in human cultures and civilization, but did not detect the difference between freely moving units and rigidly organized units (see Kroeber [11]). We have said before (Chap. 5) that in many biological systems, as with motor cars, it is eventually cheaper to reproduce and replace than to repair; hence natural selective mechanisms operate with the phenomenon of senescence and death. Thus in a system of species or of people the parts die, but the overall systems like the forests and the seas do not themselves fall if their energy flows continue, for their parts can be replaced as needed.

The theories of the rise and fall of civilizations, like the theories of the evolution of biological stocks, are often discussed in terms of qualitative factors. Whatever the mechanisms may have been, the rise and fall of a manifestation of order required large power flows in the making and in the maintenance. Any theory of history must be consistent with the his-tory of power injection into human cultures. Some of the flushes of more recent progress can be readily traced to the new sources of power, and certainly the demise of civilizations was accompanied by the loss of formerly concentrated power flows, whatever may have been the cause. White [16] describes an energy theory of culture.

Kroeber [12] related genius of innovation as a principal feature of development in civilizations. The high energetic cost of creative innova-tion related to the amount of choice and choosers was already explained with Figure 5–6. Thus innovation follows the excess power levels.

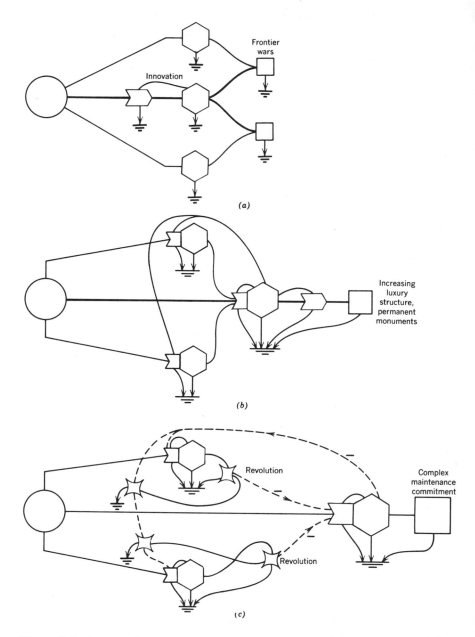

Figure 7–7 Energy diagrams representing three stages of history suggested by theory of Toynbee: (a) period of excess power; (b) period of constant power; (c) period of eroded power. Dashed lines indicate pathways which are lost.

Among the many points made by Toynbee about the rise and fall of civilizations, one concerned the development of innovations that permitted the control and incorporation of large areas. In time, stability developed without new gains within dominant civilization, and there was a catching up of the formerly less developed subordinate components, causing energy drain from and disintegration of central control.

The energetics of such a process may be illustrated by the three stages shown in Figure 7–7. In the first, power budgets are focused and concentrated through a centralized control, and excess power for organization is made available by some innovation giving a monopoly in power control. In the second stage a fully developed civilization has increased its maintenance requirements to equal the available power budget. Finally, in the decline the relative rise of control abilities in the adjacent and formerly component systems takes away some of the power flows that permitted concentrated energy for organization and specialization. Thus control disintegrates for lack of power for organization. The pieces and parts, however, are reusable in new organizations whenever or wherever some mechanism for collecting power for this purpose develops. The energetic extension of the Toynbee theory of history may be better expressed as three stages.

Our leading civilization may be in the first stage. There are vast excesses of new power, and large volumes of it go into organized innovation (science and technology). The input of fossil fuel and atomic energies to peripheral systems as they catch up is also vast and rapid. The leading civilizations build up additional structure, but at the same time the power costs of general maintenance reduce their margins of flexibility. Stage two has begun. If the power support diminishes as fossil fuels diminish, if some of the presently available power inputs become diverted from existing civilizations, or even if newly developed countries develop faster than new energy sources are found, stage three may set in because there will be inadequate power for all efforts at organization. Some operations must be reduced or disintegration may follow. The disintegration observed with the withdrawal of power flows from some African colonial organizations is a recent example. As the colonies became developed enough to gain control of power that was formerly centralized, the system disintegrated, especially when the process of disassembly lowered the fossil-fuel inputs.

The systems of nature such as the forest build up their structure until their maintenance and organizational costs equal their power budgets and disintegrate when their power sources fail. The concentrations of power in reefs, in the river rapids, and in the zones of ocean upwelling are somewhat comparable superorganizations based on the convergence of power flows, and they are vulnerable to the diversion of these flows.

When pollutions or other stresses are applied to complex natural systems, more energy is required for individual survival, diverting power that was formerly available for central organization and specialization. Such disturbed systems disintegrate with loss of specialists, and succession must start again with the generalized species.

As indicated in our discussion of energy and evolution in Chapter 5, innovation for an entirely new civilization requires that energies be directed into the process of choice and selection. Such power flows are a part of the industrialized societies, allowing them to stay in the forefront. Shocks such as the sputnik renaissance of attention to education serve at present to counteract the conservatism of complex and organized societies that Toynbee indicates was a factor in the loss of power control in twenty-seven earlier civilizations.

THE ROMAN EMPIRE

In examining designs for world energetic and economic networks, we may consider the example of the long-surviving Roman Empire which was the greatest world order ever built on solar energy alone.

Figure 7–8 is a much compartmentalized diagram of the Roman Empire running on solar energy. With innovations in organization, military mechanisms, and governmental service, a position loop system developed with enough energies converging into organizational work to produce world order. The principal energy flows beyond local provincial subsistence were the grain levies from provinces such as Africa, whereas other provinces contributed such things as slaves, recruits for legions, manufactured goods, literary services. If this system functioned like others discussed in this book, the feedback work must have been about 10 percent in order to provide loop reinforcement to Roman authority after amplification in the solar transformers of the provinces. The economic system was partly organized. Money loops were used in the smaller transactions, whereas grain and slave levies were used for the main flows, being loop-reinforced by the Roman governmental work necessary to insure these levies.

The war work of the legions expended considerable energy in preventing the surrounding provinces from draining energies from the main Roman provinces. Some of the energy supports for legion forces were drawn from the provinces upon which war work was done.

The fall of the Roman Empire continues to fascinate historians and

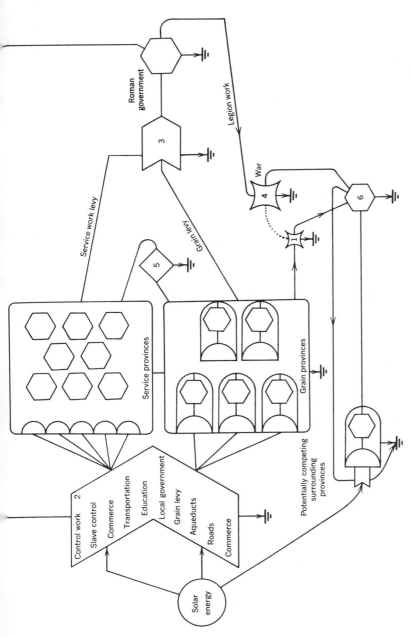

Figure 7-8 Energy circuit diagram for some principal compartments of the Roman Empire, a complex human network operating on solar energy. (1) With spread of information on organization and military methods, competing organizer countries begin to draw from some of the input energy flows. (2) Developing controls. (3) Service work of the provinces supplies slaves, legion recruits, and special manufactured goods. (4) Military work by Roman legions prevents system drains into surrounding alternative systems by blocking the work of group 6 at switch 4. (5) Intraprovincial commerce uses the currency system. (6) Surrounding provinces have potentially competing governments.

233

scientists. Examining the energy diagram, we may consider the various possibilities. If, for example, a shift in rainfall belts lowered grain production of North Africa, enough energy flow might have been cut off to lower the integrative work of the government below the loop reinforcement thresholds. Since organization is the endmost part of energy flow, it may be the first item to lose function with loss of energy.

Another explanation that can also be considered in energy terms is Toynbee's theory of fragmentation associated with loss of monopoly on control innovation. Consider again the flows shown in Figure 7–7. If part of the early Roman success depended on their innovations, the spread of this knowledge might have allowed surrounding provinces to organize and begin to take some lateral energy flow formerly directed toward Rome. The legion work costs per unit effect may thus have increased. New external energy plus losses of some previous inputs to the system, if it was already fully organized without energy reserves, may have cut off energies to governmental service work so that loop reinforcement declined followed by fragmentation. Exactly what combination of changes broke the loop system may not be known, but there can be no doubt that the flow of potential energy through agricultural production into organized services and finally into the loopback of government and defense declined. Energy diagram analysis provides a new language for the study of some old theories. As we shall see in Chapter 9, such systems can be studied experimentally with electrical simulation. It may be time for historians to move into this experimental phase.

THE MODERN SYSTEMS

Many questions arise concerning our modern industrial nations, their political structure, and their adaptation for survival. What is the ultimate solution for a country like the United States which has based its security in recent years on being ahead in development? As we have indicated in Chapter 6, the United States must help the others catch up for whatever danger this may involve, stand idly by and see others catch up without help and in a manner likely to threaten United States security, or foster a one-world economy in which the position of the United States as principal control may be submerged, a consequence not without risk when the world system may not be immediately stable. The Communist solution is to incorporate as many countries as possible.

As long as a civilization leads in innovations, it has an edge in marshaling and controlling energy flows and thus can provide energies as needed to build and maintain ever larger and more complex structure and world order (see Fig. 7–5). If, however, external countries begin to catch up and gain control over some of the energy flows formerly used to support the complex structure and world order, the organization of the dominant may fall apart, since energies for population survival tend to have priority over energies for social order. Political and military forces also depend on energy control. If the complex dominant is already spending all its fuels on maintenance without ever collecting reserves, any decline in input may cause system disintegration as discussed by Toynbee. The rising cost of maintenance that goes with progressive building of new institutions must never be allowed to cross the input budget. Means for cutting back structure are required if fuel flows ever diminish. Perhaps the present system of maintaining superior progress through research can serve as a holding action until the lacework of world economics is able to develop a single world network that eliminates hazardous competitions.

REFERENCES

[1] Ardrey, R., *The Territorial Imperative,* Antheneum, New York, 1966.

[2] Berne, E., *Games People Play,* Grove Press, New York, 1964.

[3] Cambel, A. B., *Energy R & D and National Progress,* Interdepartmental Energy Study, Washington, D.C., 1964.

[4] Cartwright, D., *Studies in Social Power,* University of Michigan, Ann Arbor, Michigan, 1959.

[5] Christy, Francis T., Jr. and A. Scott, *The Common Wealth in Ocean Fisheries,* Johns Hopkins Press, Baltimore, Md., 1965.

[6] Easton, D., *A Systems Analysis of Political Life,* John Wiley and Sons, New York, 1965.

[7] French, J. R. P., "A Formal Theory of Social Power," *Psychol. Rev.,* **63**, 181–194 (1956).

[8] Hoebel, E. A., *The Law of Primitive Men,* Harvard University Press, Cambridge, 1961.

[9] Hutchinson, G. E., "The Biogeochemistry of the Terrestrial Atmosphere," in *The Earth as a Planet,* G. P. Kuiper (ed.), University of Chicago Press, Chicago, 1954.

[10] Kowallik, W., "Chlorophyll-Independent Photochemistry in Algae," in *Energy Conversion by the Photosynthetic Apparatus,* Brookhaven Symposia in Biology, No. 19 (1967); pp. 467–477.

[11] Kroeber, A. L., *Style and Civilization,* University of California Press, Berkeley, Calif., 1963.

[12] Kroeber, A. L., *An Anthropologist Looks at History,* University of California Press, Berkeley, Calif., 1963.

[13] Lasswell, H. D., and A. Kaplan, *Power and Society,* Yale University Press, New Haven, Conn., 1950.

[14] Odum, E. P., *Ecology,* Holt, Rinehart and Winston, New York, 1963.

[15] Sellers, W. D., *Physical Climatology,* The University of Chicago Press, Chicago and London, 1965.

[16] White, L. H., *The Evolution of Culture,* McGraw-Hill, New York, 1959.

8

ENERGETIC BASIS FOR RELIGION

Any surviving network distributes power within itself as required for further survival.[1] Even systems with equal potential energy resources may have unequal possibilities of surviving if essential functions are different in their design. Well known in anthropological studies is the nearly universal presence of religion in human networks. Strong morality apparently has a survival role in programming power budgets.

Systems with programs of morality, religion, and ethics can focus and unite dispersed power resources of individuals as needed for group protection and unified actions. Especially when human societies must exist with famine, war, rapid change, and disrupted central governments, a strong religion provides a flexible focus of power, at times in individual works and as needed later in group action. The cement of a system of man may be the self-derived impetus to join into a network based on some common principles. The property of an individual as a plug-in unit is illustrated in Figure 8–1. In addition to material and energy connections required, the need for single humans to couple their behavioral inputs and outputs into a system of others is an established property of good mental health.

One of the burning questions of the modern world involves the breakdown of some past religions and moral customs and the fears that survival is thus threatened. The real question is whether these functions are being lost or whether they are being displaced by new systems for power switching that are as good or better for survival in new conditions. Con-

---

[1] Among those after Darwin who have developed the theme of natural selection at the group level of human society is Keith [2].

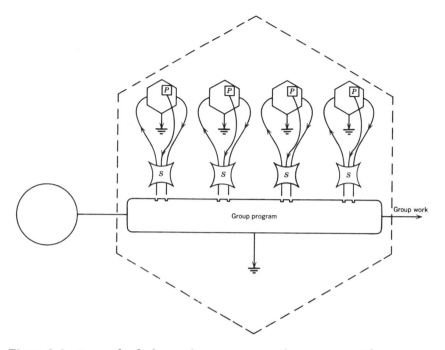

Figure 8–1 Four individuals in a human system with a master switching program control of energy flows. A minimum of one energy income line and one work expenditure line is required to plug in each of the four system compartments. The switching controls (S) depend on individual behavior programs (P) as well as the group machinery for implementing changes in flows. The flow of work from the group program is available for direction into group projects. See detail in Fig. 8–3.

sider the nature of religion in the energy network diagram, especially in its role as coordinator of switching controls.

SELF-SWITCHING POPULATIONS

In the tropical rain forest swarms of army ants alternately spread through the forest as individuals, then switch in behavior and converge their energies into common functions, exerting work flows vastly greater than those they might achieve separately. They build houses out of their own bodies or devastate a food mass. Such a network has great flexibility because of the self-switching action of the individuals, Figure 8–2 has each individual in two modules, separate operation and group operation, and a system of synchronized switches to shift from one to the other.

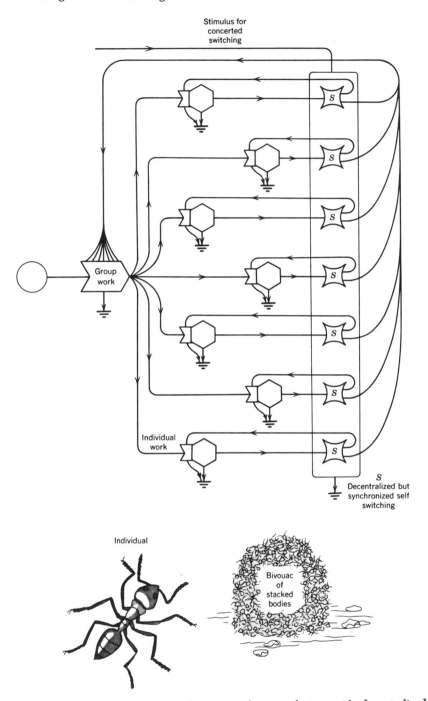

Figure 8–2 Network organization of ants or other populations with decentralized switching (S) which can provide either a common output of work or separate power expenditures.

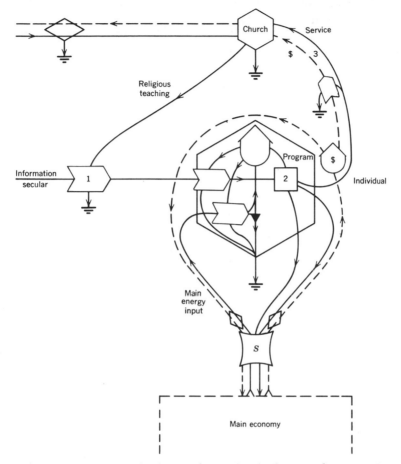

Figure 8-3 System of an individual and his church showing the processing of religious teaching and other information (1) to form a basic behavior ethical program (2) which includes his future support of the church (3) which closes the loop of reinforcement. The individual's regular service, food, and money exchanges with the economy are at the bottom mediated by a controlling switch (S) action operated by the ethical program.

In a vastly more complex way people also have means for distributing their work energies in dispersed activities until some common need causes their basic programs to serve as a decentralized master control system, focusing their work to meet emergencies, achieve unity, and perform swift group actions. The switch actions that take priority in directing the energy flows are united by common morality, and the master control systems are organized as religions. The switching actions are diagramed in Figures 8–1 and 8–2.

In the two preceding chapters the control systems of economic and political patterns were presented in network terms as social inventions evolved with man's emerging culture. Taking more priority, however, are the religious aspects of the individual. Although many of the moral functions are now taken over by the new network developments, the individual morality provides a master control and switching system when it is invoked. By diverging or converging energy flows, the religious controls as a group serve as power transformers (Fig. 8–1). The power-transforming mechanisms of animal behavior may be energetic precursors of the religious programs of man (Fig. 8–2).

Human beings require self-switching actions to place themselves into the complex networks of different societal roles at different times. Ultimately the decisions fall upon their internal moral programming, this being based on their teachings from the religious institutions or the family. In the religious system of an individual (Fig. 8–3) because of faith he contributes work to the church and receives the work of its teaching. These mechanisms provide a program of control for the switching actions. The individual will only plug into a network of functions, which is consistent with his ethical teachings.

THE RELIGIOUS WORK LOOP

Figure 8–4 is an energy network diagram summarizing the circular loop of work processes that can generate and maintain adaptation of morality and religious institutions. The individual has faith in the social system and therefore switches in his component energies. Of the many variable flows and actions resulting, those that develop loops need to be retained and programmed for the future. It requires a prophet and a continuing institution of spokesmen to see what these are and put them in the symbolic and vernacular language for the times, in effect translating data from the network. The religious institution and spokesmen, after performing cascading work, provide what is necessary in training and indoctrination to help people, especially the young, see the pattern, adopt the faith, and accept the symbols of the religious package and the dogmas of what is good. The circle of work provides a circle of consistency for the definition of revealed truth. What has been energetically system-rewarded is taught as religious truth to strengthen faith, for it leads to further reinforcement and helps the network survive.

Faith in a religion that supports a way of life is necessary to the group's

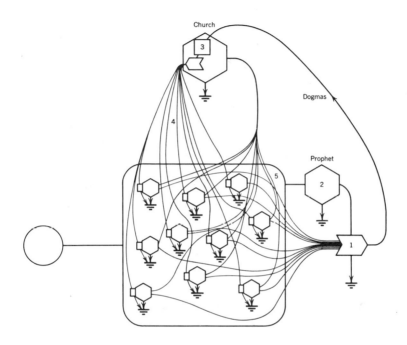

Figure 8–4 Energy diagram of religious control loop for a social system. Religious work (1) of individuals coordinated by a prophet (2) develops program of religious dogmas (3) which favor the stable energy flow of the network and become the bases for the church. The church provides religious indoctrination (4) and receives the support of the individuals in the social system (5).

coherent survival and continued efficiency. The need and importance of religion thus have a hard energetic basis. The loop through the prophet may be necessary to keep religion adapted to changing times. When there is no mechanism for change, one religion may be replaced by another.

According to the loop reward mechanisms of self-design, experimentally tried controls that cause more circular energy flow receive reinforcement and tend to be established. Religions as energy control programs may develop competitively in this way, as suggested in Figure 8–5 which shows two religious systems operating together. Each religious system is conservative because it has its own loop reinforcement of support and teaching, but the winning of additional support and needed inputs may depend on its effectiveness as a control in fitting individuals to the main system of the times. The faith that engenders a better energy flow for its believers receives energy reinforcement to direct energies (or energy-stimulating money) into the religious institutions. Believers in the older (or newer) religion shown in Figure 8–5 are not receiving much energy reinforcement from the main system. This lesser system may not be giving much loop reinforcement if conditions have changed. For example, a dogma against evolution may prevent a person from receiving a scientific education and cause him to be less able to work at a modern occupation; thus he cannot provide his group with the reinforcement work needed to expand its realms of influence.

PRINCIPLE OF COMPLEMENTING CONTROL BUDGETS

In evolving complex control systems the simple loops and switching mechanisms become specialized into many economic, political, religious, and other types of programs, all working on the same population units. As we have seen in previous discussions, these various controls have quite similar network energy diagrams. Energetics supplies a basic constraint, that the total sum of the control systems cannot exceed the energies available for control. Thus a system with strong religious controls may need fewer political institutions and laws. Systems without money may require a greater development of political and religious institutions. Energy-rich systems can support many more control systems.

NETWORK BASIS FOR UNIVERSAL GOOD

Since many power networks have common properties, it follows that the programs of religious switching control, although they have developed independently, may perform similar services and have like dogmas of

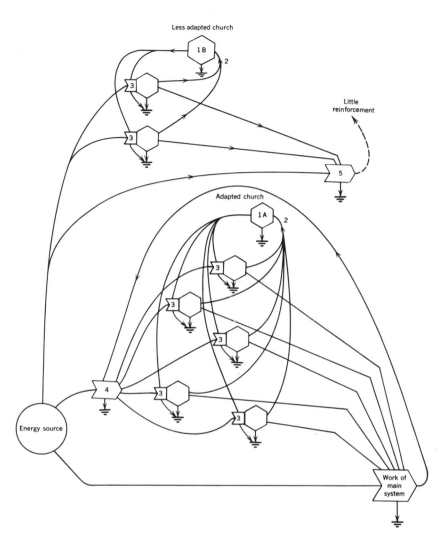

Figure 8–5 Energy diagram for two religious groups in the same economy with one receiving strong loop reinforcement through the adaptability of its program in fitting its members for best group functions. (1A) Dominant religion, members adapted to system work; (1B) secondary religion, members not adapted to system work since there is no reward loop from work done; (2) energy support to the religious institution from its members; (3) work of the religious institution in developing the individual's faith and morality; (4) compartmentalized representation of the loopback of control and reward work from the main system to the group supplying more participation and adaptable service; (5) work of less adapted group without a loop reinforcement.

Table 8–1   Ten Commandments of the Energy Ethic for
Survival of Man in Nature

---

1. Thou shall not waste potential energy.
2. Thou shall know what is right by its part in survival of thy system.
3. Thou shall do unto others as best benefits the energy flows of thy system.
4. Thou shall revel in thy systems work rejoicing in happiness that only finds thee in this good service.
5. Thou shall treasure the other life of thy natural system as thine own, for only together shall thee all survive.
6. Thou shall judge value by the energies spent, the energies stored, and the energy flow which is possible, turning not to the incomplete measure of money.
7. Thou shall not unnecessarily cultivate high power, for error, destruction, noise, and excess vigilence are its evil wastes.
8. Thou shall not take from man or nature without returning service of equal value, for only then are thee one.
9. Thou shall treasure thy heritage of information, and in the uniqueness of thy good works and complex roles will thy system reap that which is new and immortal in thee.
10. Thou must find in thy religion, stability over growth, organization over competition, diversity over uniformity, system over self, and survival process over individual peace.

---

what is good, explainable in energy network terms. In Table 8–1 are listed some properties that have generated loop reinforcement in the past, thus continuing themselves as network features of surviving religions. With the aim of relating the ethics of energy to modern needs, the properties are given in the form of commandments. Perhaps statements such as these can hasten the evolution of a relevant religious dogma.

EMERGENT PROPERTIES OF NETWORKS

When a person is part of a system, he cannot easily see what his role accomplishes. A man in an auto assembly line who has never viewed the whole plant will not understand what his work accomplishes. Unless he understands the system thoroughly, he will not have any inkling of the network of controls that may or may not exist to keep the flow of cars continuous, adapted to inputs, adapted to outside demands, and stabilized in the face of fluctuations.

With the fantastic complexity of the modern network[2] of people and

---

[2] The Vernadsky noosphere is defined as realms of phenomena so dominated by the activities of man that they are power-determining. A noosphere is possible only

groups, each person understands perhaps his local situation but not what he accomplishes for the group and how his accomplishment is controlled. A nerve cell in the brain does not know what thoughts are. It only feels the passage of slugs of electric waves, the flows of individual fuel, the outflows of wastes, and perhaps the impact of control messages from hormones and cellular regulatory circuits. The individual's role in his own institutional networks is similar. It is a principle that no system can understand itself since understanding requires more units than operation. The network of society has a form of thinking since it is a computer network in itself, quite over and above those it encompasses—the electronic computers of industry and the even more remarkable computer, the mind of man. The network is doing things to keep itself regulated, adapted, and consistent with energetic laws that the individual cannot envision. Some people not knowing what it is all about fear that all is disintegrating. Others develop a new faith in the wonders of the network.

If the supernetwork of man, machines, and nature has its own centralized group power control systems, including patterns for inheritance and transmission to the future, a group morality, a property of the supernetwork not even visible to man, must be emerging. Just as with any computer network, we can find ways to portray to humans what goes on if we can program it with special plotting screens of perspective. One of the tasks of social science is acquiring means for communicating with the supersystem by finding its language. Does the supernetwork have a self-consciousness?

A FAITH FOR EACH NETWORK, NEW GOD EMERGING

When man was a tiny part of the stable complex forest, his faith was in an umbrella-like energy system with God identified as the intelligence within the mechanisms of forest control, the system. Primitive forest peoples such as the early Druids of Europe had religious faith in the forest as a network of gods operating with intelligence. A stable forest actually is a system of compartments with networks, flows, and logic circuits that do constitute a form of intelligence beyond that of its individual humans.

When man entered the sharply pulsing climates, the erratic alternation of favorable energy flows and famine were capricious controls. The faith

where and when the power flows of man, or those completely controlled by him, displace those of nature. This kind of dominance over the power of nature is now prevalent in industrialized areas, but these areas survive only because of the purifying stability of the greater areas of the globe not yet so invaded and polluted.

that developed entertained miracles, infinite patience with catastrophe, and other adaptations to an energy control that came mainly from outside man's system. Story [4] found the Bushmen of southern Africa with intimate knowledge of its natural dry ecosystem serving to integrate and manipulate many of the processes.

Now with new complex and stable networks, partly of man's own making, a new kind of energy system provides a different umbrella. The new religion emerging is a faith in the new control system that provides rewards to those individuals who receive enough education to participate. The faith of modern man in the magic of his new supporting network is not so different from the faith of the primitives in their forest network. Possibly consistent with emergent properties of networks as the essence of adaptive God is the thesis of Northrop [1]. In his Chapter 1 he notes the creativity of man's neural network and also finds "nature creative," the source being called "evolution, God, Allah, Yahweh, Brahmen, Nirvana, or Tao."

In the modern world faith in our network group systems, in science and in big institutions, is receiving reinforcement, although these faiths are not yet considered religious. Older faiths still serve for those who are not receiving much energy reinforcement from the new systems or for those who are conservative about redesign. There is no single prophet; the prophetic function is instead dispersed over the educational system through which the ethics of the new system is being taught. Because systems are changing so fast, there is not time for one pattern to develop the status of a religion before changing again. Thus a void is created. Many young people, seeking a workable faith for the new patterns, do not find it among the religious institutions where they search, for it is evolving elsewhere and is still hidden. The time may be right for another prophet if a plateau of diminished change develops. The writings of new religious spokesmen, for example de Chardin [1], contain such concepts as a system psyche, consciousness of the system, and love as measurable by energy, noogenesis against the entropy stream.

Much of the conflict of the liberal and conservative concepts in national politics concerns the relative degree of trust in the new and experimental power controls of the centralizing society and in the older religious morality time-tested for simple worlds. The old ones may not be safe under our new conditions.

The history of religion provides some comparable situations in which the establishment of new kinds of social energy networks were followed by the emergence of new prophets and religions, the new dogma taken as revelation from the network design itself. Thus Christ arose in Roman times, Joseph Smith in special frontier times, and the communist religious dogmas in the new Russian pattern.

If God is defined as the source of revealed truth and if such truth is defined by the complex network of man and nature in terms of its own survival (see the prophet loop in Fig. 8–4), God becomes identifiable with the networks of which individuals are mere parts. As we have seen in previous discussions, such networks may have superintelligence, control, protection of its parts, a means of arriving at good, a truth definition mechanism, and energetic reality justifying faith. In the energetically rich twentieth century, networks of ever-greater complexity are emerging. In other words, God is growing and is emerging in new forms, but some of God's older forms may have to be stored away until the simpler networks return, if they ever do.

THE INDIVIDUAL AND THE SOUL

Inherent in any overall consideration of network survival is the value of people to survival and the relation of such a concept to democracy. As a person passes from the fertilized egg through birth, childhood, schooling, and early job experiences, his internal network may develop greater order, more memory storages, and more of the abilities that serve the system well at the individual level, the family level, the local level, or the worldwide levels of action in intellectual or political affairs. The energy cost of all this improvement is the work spent by him and on him. Operating against the quality-generating work are the error-generating actions that must be balanced by maintenance action. Eventually in old age and in death the person as a unit system loses some of his stored order or through some irreparable change loses much of his ability to serve. If the soul is identified with a level of special complexity of the unit system, it emerges during network formation and its unique aspects become incorporated into the total system's information storages and inheritance as a form of immortality.

To a considerable extent, the power control roles of people grow when abilities are increased and strengthened through various career sequences involving seniorities, recognitions, and judgments by others. The structure of democracy provides for each person to be brought forward for possible loop reinforcements which may or may not develop and ultimately bring that person greater growth and more important roles. Because the system is realistic, the right to start such a route is not also a right to arrive, except through step-by-step development along a line that turns out to be valuable to some part of the system and thus provides loop reward reinforcement.

Another feature of the democratic system is an equal vote in the control system determining top government. The government thus has equal input control from those who are and those who are not educated or otherwise loop-reinforced. These equal votes may serve some very useful system functions including conservatism, assertion of governmental effort for the system's weak members, and retention of control by generalists rather than by specialists whose judgment may be one-sided. Usually the rights of democracy are stated as absolute and self-evident, whereas their real justification may be in the values that such functions have for system continuance and survival. If the network is identifiable as God, then such network needs are God-given truths.

THE FALSE GODS OF POWER, THE AUTOMOBILE

It is probably understandable and reasonable that power is important in the programming of the surviving cultures. The love of power, like sexual drives and hunger, are subroutines that have their function, when called in to action, properly balanced in the whole pattern of an adaptive behavior. Like these other packaged drives, they become lusts and destructive only when they take an improper percentage of the system's power. The old religions provided warnings of most of the old dangers of diversion and designated the excesses as sinful, as dissipation. In energetic language, they were useless diversions of valuable potential energy flows. In religious language the admonitions were to avoid false gods that would lead the enterprises into waste and the individuals and cultures into destruction.

Great quantities of cheap high-grade energies have not been generally available before recent times so that the lust for power could not take excessive dimensions. Our new forms of excess were not included in the old religions. Now, however, the largest sources of dissipation of potential energies hardly related to the economy of a stable surviving system are the automobiles, the love of automotive power, and the fantastic structures of concrete that stand as idols to the false God of Power-Lust. The love of speed and acceleration takes hold of all of us, draws the first buying power of the adolescent, dominates the thinking of his elders in the budgeting of funds, and pumps more and more individual fossil fuels into accelerations and speed. The luxury automobile and highways are surely the dominant part of our culture that is unecessary, destructive of human cultural coalescence by injecting large random energies, keeping people on the move and out of more local group operations. (Table 2–1 shows the enormous difference in power levels of even a small automobile compared to the other units of our biosphere.)

Because the kinetic energy of a moving object goes up as the square of the velocity, acceleration followed by braking leads to much greater energy costs of transportation than a slow steady speed against steady friction (see Chap. 2). For any process of needed movement there is a speed that is too slow so that competing activities will displace it. There is also a speed that is too fast, dissipating so much energy unnecessarily that the unit loses out in competition because of its waste preventing it from doing what it could. There is an optimum speed for processes for survival [4]. Our Western culture is clearly in the excessive speed and is temporarily getting away with it because of the excessive temporary fossil fuel availability. This situation becomes lethal as soon as there are competing systems that use energies better or the energies dwindle. Our sin of excess power dissipation has displaced older weaknesses that are proportionately minor now. The church doctrines move to include the new sins but they can hardly do it with the old revealed dogma that has no mention of automobiles, highway folly and giant airplanes.

When anthropologists look at the fossil remnants of past civilizations, they often find large structures like the pyramids which were of great significance to the past civilization in requiring a high proportion of energy, but which in our present culture seem unimportant to survival. Future cultures looking at our cultural remnants may find gaunt, empty expressways. These concrete fantasies may even develop religious significance to less energized cultures following in the same way that Roman cultural inheritance achieved religious significance in the Medieval Ages. Symbols of a past high energy era may imply past magnificence of god to a less energized later period. Are civilizations known by their gods of survival or their false gods of excess energy?

INCLUDING THE NATURAL SECTOR IN OUR
RELIGIOUS SYSTEM

Named the Land Ethic by A. Leopold and well stated by White [6] is the error of our recent religious faith in excluding the life of nature and its system from our religious ethic. Whereas many of the earlier religions kept man in mutual allegiance with his life support, our fast development of urban culture on fossil fuels has moved further away from duty to nature as to oneself. Few now would regard the cutting of a tree or the draining of a swamp as evil, but facts of the ecosystem require that ethic again. We may accelerate this change by putting energy values on the natural sector and from these showing its value to those used to thinking only in money terms. Such calculations show nature more valuable than man.

ENERGY AND VALUE

The storage of order from previous disorder constitutes a lowered entropy and also a storage of potential energy since the improbable state can go back to the probable state. The energy storage comes about in setting up the order and fixing on one from among many possible combinations. If we now move to substitute one definite arrangement for the other, both having the same information content and potential energy, we spend organizational work without creating more storage. But in so doing we may have changed from an arrangement that was of little value to the system to one that is very important. Both patterns may disintegrate to the same extent, and both may cost the same to maintain. Both may have the same probability and information content. They may differ, however, in the energy it took in accumulated work to find the plan, the useless one arrived at perhaps without energy cost, the other requiring the immense energy expenditures of creative work process (Fig. 5–6). The useful one has amplifier value in guiding other flows, receives loop reinforcement, and is maintained, whereas the other erodes away. The energy value in structure has three parameters, one the potential energy storage values, second the work cost of replacement, and third the energy value of flows it can control through work gate actions.

If souls and moralities are emergent properties of complex self-maintaining networks, the energy values of the cumulative work may be vast, derived from the earlier work investments of generations. In the lesser organisms and ecological networks we may also find great values, although the customs of man often draw a sharp line that sets him apart from other creatures and energies.

ENERGY TEST OF EVANGELISM

The broad energy ethic leads us to some different conclusions about trends of our times and the needs of our youth from those of our leading evangelists whose personifications and language sometimes lose them in the new wilderness. For example, sexual behavior excesses which were seriously deterimental to the scarce energies of a primitive culture, become irrelevantly small in a power-rich system where the great sins are in the wastes of enormous calories in the destruction of the life-support system, in the useless worship of the false gods of the automobile, and the greed of power expansion at the expense of those yet to come. The religious leaders who are trying to adapt need to pose questions as to which older teachings are now evil because of their energy distortion. In times and places of erratic energy, the pattern of fitting man to a role of programming the environmental system requires focus of energy into births, these being balanced by microbial controls automatically. How right is this birth emphasis in our present energy regime? Which is the evil—to give food and interrupt the development of a starvation preventing food-work system or to withhold charity, helping a hand instead to a job?

Often religious preachings call for personal covenants with God as a spiritual separation and retreat from the practical world. Yet, if God is the great system, in which will salvation be found?

In medieval times the preservation of the information of an earlier, energy-rich classical period of information development was critical to the renaissance of a high order of man's flowering. It was appropriate to give oneself to God as the network by entering a monastic refuge to aid the retention of knowledge. But now, again in energy-rich times, who gives himself most to God, he that retires into a backwater with reapplication of biblical language awaiting a coming of God or he who dedicates his soul to the stabilization of the great system of order exploding miraculously on earth now through the emergence of the fossil fuels, truly already a new coming of God and the millennia. God is in exponential growth.

ANGELS, DEVILS, AND THE ENERGY LAWS

Whereas placing religious ideas, rules, and institutions in the measurable language of energies helps put science to work for religion, the exciting personifications and symbols of religious teachings are the ones which reach simple children in their formative stages. Illustrated in Figure 8–6 are the angelic forces of good which may be identified with the development of order versus the inherent tendencies of disorder and destructive systems symbolized by the forces of evil, the devils. The development of revealed truth through a contribution of immortality in a selection process at the pearly gates may be compared with the network's (God's) natural selection of compatible survival formulae which then become right.

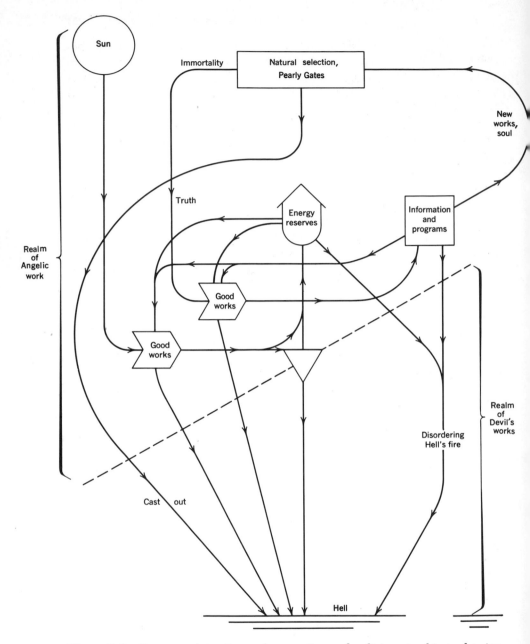

Figure 8–6 Common abstractions of energetics and religious teachings showing angelic operations of order, evolution, and selection of information above and the evil processes of disorder, dissipation, and heat death below.

252

Defined in this way in systems of accelerating power influence there must necessarily follow increased evil and disorder, but also evolution of even greater good in the surviving networks.

If there are parallel concepts in energetics and religion for the same phenomena, what a paradox that the one can be taught in schools and the other, although simpler, must remain allied to sporadic instructional efforts. What would happen if we combine the concepts (Fig. 8–6)?

SUMMARY

The key program of a surviving pattern of nature and man is a subsystem of religious teaching which follows the laws of the energy ethic. Whereas the earlier tenants of religions were based on the simple energy realms of their time, the new sources and large magnitudes of power require revisions of some of the mores and the personifications used in teaching them. We can teach the energy truths through general science in the schools and teach the love of system and its requirements of us in the changing churches. System survival makes right and the energy commandments guide the system to survival. The classical struggle between order and disorder, between angels and devils is still with us.

REFERENCES

[1] de Chardin, T., *The Phenomenon of Man*, Harper Torchbooks, Harper and Row, New York, 1961.
[2] Keith, A., *A New Theory of Human Evolution*, Watts, London, 1948.
[3] Northrop, F. S. C., *Man, Nature, and God*, Trident Press, Simon and Schuster, New York, 1962.
[4] Odum, H. T., and R. C. Pinkerton, "Time's Speed Regulator: The Optimum Efficiency for Maximum Power Output in Physical and Biological Systems," *Amer. Scientist*, 43, 331–343 (1955).
[5] Story, R., "Plant Lore of the Bushmen," in *Ecological Studies in Southern Africa*, D. H. S. Davish (ed.), Junk, The Hague, 1964, pp. 87–99.
[6] White, L., "The Historical Roots of our Ecologic Crisis," *Science*, 155, 1203–1207 (1967).

9

ELECTRICAL SIMULATION OF
ENERGY NETWORKS

Using electric networks as models for various other networks is now a
principal technique in many sciences. Perhaps a good way to summarize
the story of force pathways and energetics of man and nature is to present
the electrical energy simulator, an extension of a simple hardware device,
the passive analog, which was used for studying such simple networks as
water pipes and heat diffusion as early as 1935 [13]. Also, examples are
given of simulation by operational analogs and by digital computer
program. First, let us review the energy network language that has been
used throughout the previous chapters.

REVIEW OF THE LANGUAGE OF ENERGY FLOWS

Phenomena of man and nature are portrayed with a special language and
symbols in Figure 2–4. Both nature and society consist of units and energy
flows between them and are diagramed as networks of energy compartments (tank symbol or self-sustaining symbol) and interconnecting pathway lines representing energy flows. Any part of the system in which
energy is stored is given a compartment symbol, and the quantity of storage is indicated with the number of calories.[1]

---

[1] Since storages may rise and fall in a compartment when forces change, way of
designating the storage property of the unit (from basic physics) is to write a value
for $C = Q/X$ where Q is the quantity of energy stored when packed into the compartment by the force X. A storage pushes out with the force X in the process of im-
pressing its energy storage on later processes. For various kinds of storages (water,
electricity, chemical fuels, etc.) there are various storage forces. The thrust of a
population of forces X is N, the symbol chosen for population force. C is called

The energy stored in a compartment is packed in by forces that do storage work, creating an accumulation of potential energy. This potential energy exerts its ability to do work through the outward forces that follow the flow paths and arrows of the energy diagrams. Any force has an equal and opposing force, most often frictional.

Since all potential energy goes into dispersed heat while doing its work, all flows are terminated in a heat ground symbol indicating that the energy that is flowing is being dispersed in the random motion of the molecules of the environment (heat).

When the flow of energy from one potential energy source does work so that another flow is proportionately aided, we indicate the work relation with the pointed block. Energy passes into heat while aiding the second flow, making the flow of the second in proportion to the product of the two flows.

Networks are drawn with these and other symbols, and we have formalized our description of systems. Probably, some readers may know some different network languages being developed such as logic circuits, signal flow diagrams, electronic wiring diagrams, operational analog computer mathematics, and digital computer flow charts. These have to do with flows of unit pulses, electrons, voltage mathematics, and switching choices. Since energy is a common denominator for expressing processes of all kinds, the network language of energy flows may be more general using laws of energies and forces to provide an inclusive point of view under which to gather together various types of languages.

PASSIVE ELECTRICAL ANALOG CIRCUIT

The passive electrical analog circuit provides for the flow of electricity as an energy carrier over the same network pattern drawn on the energy diagram of the real system. For each flow pathway there is a wire, for each storage function there is an electrical hardware unit that operates quantitatively in the same way. The chart diagram of the passive electrical analog circuit has often been used to represent nonelectrical systems because it provides a simple language whose behavior is well known and whose energy flows are analogous. Thus many plant ecologists and micrometeorologists draw equivalent circuits for the movement of gases and other materials in forests, adopting such terms as resistance for the flow

capacitance in electric systems but in ecological and human systems is the ratio of fuel value stored (Q) to the population force (N). N is proportional to the surface area available for reaction, which in animals controls the rate of use of foods in metabolism.

patterns [3, 6]. Passive analog circuits used for description are often called *equivalent circuits*.

By using energy circuit diagrams of real systems first, we are unhampered by electrical restrictions when we set out the patterns of a real nonelectric network. Having drawn such a circuit we may merely substitute the hardware using the equivalents given in Figure 9-1. The analogous quantities of the two systems are set out in Table 9-1.

THE USEFULNESS OF PASSIVE ANALOGS

One of the useful techniques in many network sciences has been representation of the network with electric wires and parts whose behavior simulates the phenomena. Analog simulation has many purposes. The process of setting up flows and functional parts is a powerful stimulant

Table 9-1   Equivalent Concepts and Definitions

General Name	Symbol	Electrical Energy Simulator
Force and population of forces	X, N	Voltage
Storage mass (with associated energy)[a]	Q	Charge, coulombs
Flow of matter (carrying energy)[a]	J	Current, amperes
Storage-force ratio[b]	$C = Q/X,\ Q/N$	Capacitance, farads
Resistance	R	Resistance, ohms
Turnover time[c]	$T = RC$	Time constant (seconds if R in ohms and C in farads)
Force-flux law	$J = \dfrac{1}{R}X,\ \dfrac{1}{R}N$	Ohm's law

[a] The ratio of energy to stored or flowing mass is different and follows different formulae depending on the type of matter. J and Q may also be pure energy such as light, sound, or other waves. In many biological, ecological, and social systems the energy is proportional to the mass of foods and fuels flowing.

[b] In biological and ecological systems, the storage-force ratio may be defined as the volume to surface ratio, which varies according to the size of the storage unit, being large in elephants and small in bacteria.

To scale the hardware circuit set a mass equivalent to charge; set the values of flow per unit time equivalent to electrical amperes. The unit of time used for the flows will be the unit in the time constant. Scale energies to flow for each pathway according to the kind of energy flowing there.

[c] The flow rate per gram is sometimes called the transfer coefficient and is the reciprocal of the turnover time. The turnover time is the time it takes to displace the storage with equal mass at steady state. It is the time required for 63% of the storage to discharge when there is no inflow.

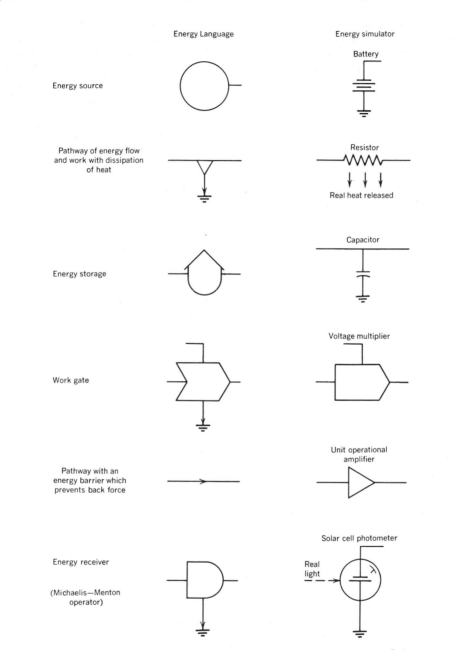

Figure 9–1 Symbols for electrical hardware items which are substituted for parts of the energy network diagram in making an energy simulator.

to the imagination, helping one to be precise, to ask new questions about the particular networks, and to secure the right measurements. The electrical flows, being very fast, can simulate much slower systems. Because electrons and their control pieces are very small, they can model large systems that would be expensive or impossible to manipulate experimentally.

For example, the flows of groundwater across the state of Nebraska depend on the inflows of rain which affect the water table, the outflows of pumping wells, and lateral flow. The real system is hundreds of miles in extent. The effect of a pumping well on the water in the vicinity was found out directly without the expense of drilling a well [4]. By making a network of conducting water pathways (simulated by resistors) and water storage units (simulated by electric capacitors) as in Figure 9–2, we may follow the consequences of rain and pumping by determining the distribution of electric voltage which represents water table height in the model. The flow of electrons in wires is so fast that fifty years of changes can be observed in a short time. The properties of the ground in conducting water are represented by using pathways (resistors) of larger or smaller electric resistance, and the porosity of the ground is indicated by using larger or smaller electric capacitors. The input flows are indicated by the battery voltages and the outflows of pumping wells by short circuits. The analogy is possible because the law of electrical flow (Ohm's law) is similar in form to the law for hydrologic flow.[2] The flow of electric current thus simulates the flow of water current, and the height of the water table is simulated by voltage.

ANALOG FOR THE ENERGY LANGUAGE

The passive electric networks can also be used to simulate and model the general flows of energy in all systems, providing some simple rules are followed. The previously discussed example of water flow in the ground across Nebraska also simulated the flows of energy. Potential energy in the water varied with its height (water table height), and thus the voltage of the electrical model served to show potential energy in the water. In this particular instance the energy in the water table was in square

[2] Ohm's law $[J = (1/R)X]$ states that the electric current of electrons J is proportional to the electric force (voltages) pushing the flow, where R is the resistance property of the wire pathway. Hydrologic law states that the steady flow of water through a frictionally resisting medium such as the ground is in proportion to the difference in water pressure (produced by the height of the water table) where R is the resistance property of the ground (see Table 9–1).

relation to the water height.³ The energy from the voltage passed into heat in the resistor pathways of the model in the same way that potential energy of water in the elevated water table passes into dispersed heat in the frictional flow through the ground.

The passive electric network serves as an energy flow model without any special devices or calculations for those systems in which the potential energy is in proportion to the causal force or population of causal forces. For the systems of man and nature most of the flows of energy are directed in proportion to the storage of energy because the flows are mostly produced by populations of organisms, people, and other units which act in parallel. The flows of food through a population are in proportion to the number of packages of food being processed, each of which has a fairly fixed calorie value of potential energy. The analogous properties of the passive network allow any of the energy network diagrams of this book to be simulated by allotting one conductor for each pathway and appropriate functional units for each compartment or function drawn.

A passive network like that shown in Figure 9–2 serves also as a general model for the flow of energy in systems of man and nature. For example, the electrons represent the recycling flow of money and materials, whereas the voltage represents inflows and dispersion of potential energy as it does its work. There can be as many branching wires as we have branching flows. Application of the passive network to ecology and to the social sciences is just beginning. It should help unify the various sciences and intertranslation of languages around the network models.

Many companies manufacture operational analog systems⁴ for use in research, teaching, and management. What is badly needed is a moderately priced passive analog system which can be purchased by universities and research laboratories for energy simulation of the many networks that fall under the umbrella of the energy flow language. Economics, sociology, military operations, food chains, and behavioral systems may

---

³ In many systems of nature, however, the energy is stored with different relations that are nonlinear to the forces packing the energy into the potential energy states. For example, the energy packed into the compressed gas of an automobile tire has a logarithmic relation to the force exerted by the compression of the air pump. The energy stored in an electric condenser is a function of the square of the amount of voltage used to impress the charge.

⁴ The quite different *operational analog computers* do a different job from the simulation just discussed. The operational analog system manipulates voltage according to a previously written mathematical equation. Thus we plug in units that add, subtract, divide, multiply, and so forth. The flows of the wires, the energies, and the resistances do not themselves simulate flows, energies, and resistances of other systems.

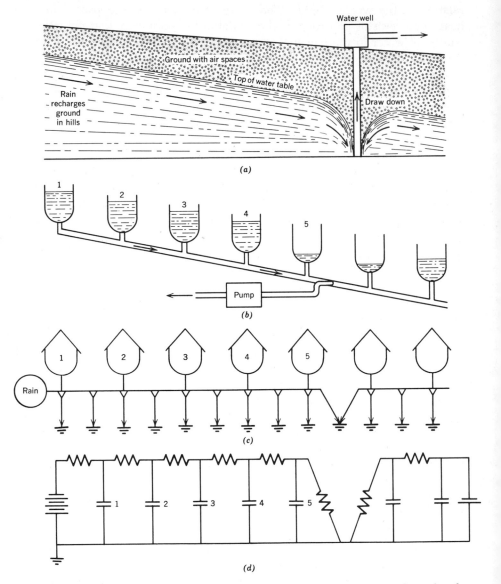

Figure 9–2 Example of a passive analog system simulating energy flows for the horizontal flow of ground water. (*a*) Sketch; (*b*) compartmental model; (*c*) energy diagram; (*d*) passive analog; voltage simulates the hydrostatic head of water. The electric current simulates water flow.

all be modeled by the energy simulator. Many of the simpler physical systems are also best modeled with the energy simulator, including the groundwater flows mentioned earlier and molecular diffusion.

The energy simulator can model all systems, even those whose energy storage functions are not linearly proportional to the force if we attach an electrical indicator that reads energy as the log of the voltage, as the square of the voltage, or in other appropriate terms relating the potential energy to the form of force that stored the energy. Can this simulator become important to the social sciences? Whereas operational analog methodology involves the writing of differential equations first, passive analog methodology bypasses the equations except to verify the similar behavior of the particular hardware pieces used. The energy network language and the electrical model are forms of mathematics in themselves, but forms that naturally resemble the normal ways of thinking in biology, ecology, and the social sciences. These sciences have been drawing flow charts and force diagrams for years; the energy language and the passive analog for energy flow are the natural extension and formalization of these established ways of thinking.

THE TRANSIENT ENERGY PULSE

Through all kinds of networks, water flows, economic flows, nerve connections, ecological food chains, and electric circuits, the start of a flow of energy under the impetus of a driving force from a potential energy source produces a wave of charge phenomena in the downstream compartments. For example, increased water level from rain on the left of Figure 9–2 would be a pulse. As shown in Figure 9–3, the ripple of energy storage and force passes from input source to one compartment after another, filling the first with the storage that is possible with the force at that point, then passing to the next compartment, and so forth. The graph of the rise of storage has a characteristic form, and the time required may be computed from the values of the pathway resistances and the storage capacities (see Figs. 1–6 and 11–1).

If the input of energy directs its force for a short period and then goes off again, it is termed a square wave (Fig. 9–3a). The transient sequence down the chain of compartments is then a series of rises and falls, the peaks arriving later and later. If the energy arrives with repeating pulses, the train of transient peaks keeps repeating. Depending on the resistances, capacitances, and time delays introduced, the system's pulse may be damped out downstream or reinforced to produce an oscillation. Thus populations, economies, and radio networks may oscillate or be

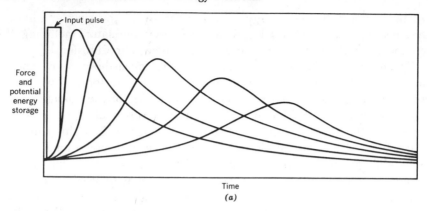

Figure 9–3   Examples of passage of an energy pulse producing transient phenomena. (*a*) In a chain of five compartments in hydrological, ecological, passive electrical, and other comparable systems, such as compartments numbers 1–5 in Figure 9–2. (*b*) Passage of radioactive cesium tracer through a forest ecosystem [11].

smoothed. The whole advanced science of electronics dealing with the transient phenomena of electron flows becomes applicable to other networks and to the ecological and social networks as well.

The charging wave that stores energy in a chain of compartments is related to, but different from, the charging wave of a tracer, such as a radioactive element in a mineral cycle or a body of marked dollar bills in an economic system, for these pass in small amounts. In filling a storage (growth) compartment the energy flow builds up population force and thus changes the flow rates. The tracer, however, disperses itself along with other units in a steady-state flow without changing the amounts of storage, the forces, or the flow rates of the material being traced. See the example in Fig. 9.3(*b*).

ANALOG OF THE BALANCED ECOSYSTEM

If a downstream compartment does necessary work on an upstream compartment, the system operates by loopback and one compartment's flow must wait on that of the other, even though energy passes by outside unused. Figure 9–4 shows circuits and analogs for a balanced ecosystem, either of the world or of the small closed laboratory microcosm, with *P* and *R* in balance, the same as first considered in Figure 1–5. An electrical analog simulator is given in Figure 9–4(*d*). The electrons represent the flow of the main carbon cycle. The voltage represents the storage of energy during photosynthesis and its subsequent decline in driving the

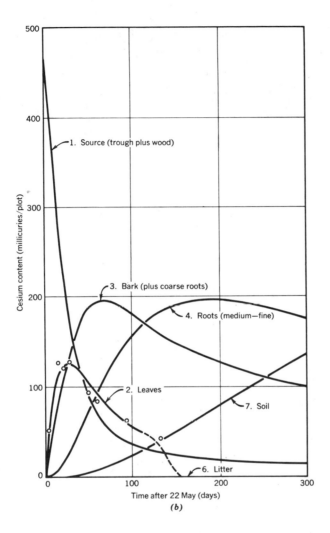

Cesium content (millicuries/plot)

1. Source (trough plus wood)

3. Bark (plus coarse roots)

4. Roots (medium—fine)

2. Leaves

7. Soil

6. Litter

Time after 22 May (days)

(b)

consumer food chains that do the necessary work for survival of the system. As shown earlier in the Figure 1–6, the charge-discharge occurs with each pulse of the energy of the day or season, and the metabolism of one day is geared to the activity of the next night so that the two stay in phase.

The longer the time between energy pulses, the larger the storages must be for the system to remain alive. Thus, for stable flow storage modules are designed partly to extend the flow between energy pulses.

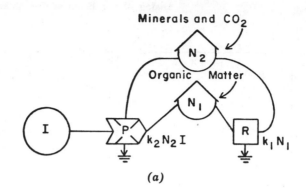

(a)

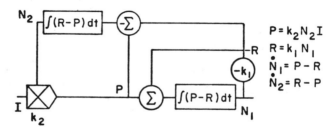

(b)

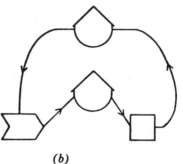

$$P = k_2 N_2 I$$
$$R = k_1 N_1$$
$$\dot{N}_1 = P - R$$
$$\dot{N}_2 = R - P$$

(c)

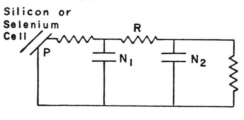

(d)

TWO-POPULATION LOOP

The balanced system in Figure 9–4 is a two-compartment loop. In the positive reward feedback that a downstream population such as man practices on an upstream component such as his crops or cattle, a relatively small amount of energy goes into the work of directing a ten- or hundredfold flow toward himself. The relatively small drain, about 10 percent, that he puts on the upstream system may put the upstream population at a competitive disadvantage, but his feedback of 3 percent of that compensates for 10 percent drain because his feedback is in effect amplified. Closed loops thus reward all components and provide for amplification and stabilization.

Two such populations in sequence serve as successive storage summations of food inputs. If a pulse runs through them, it takes time to fill up the first population storage before the flow of increased food can pass on to the second. By the time the downstream unit is filling, the upstream stock is declining. Then the rise of downstream stocks, although less in magnitude, through their amplification effect in feedback control may stimulate the upstream growth by supplying outside energy again. Thus, depending on the storage quantities and flow rates, the system constitutes a two-compartment oscillator system or it dampens out the pulses of the input energies. Closed microcosms, normally adapted to a night and day program, continue to show some sign of oscillation for a day or two after the input pulse stops, possibly because of programs in living cells.

COMPARISON OF FLOW AND MATHEMATICAL LANGUAGES

The example of the circular mineral cycle of the closed ecological system (or the world biosphere) can be used to show that different network language explains the same phenomena in different ways. In Figure 9–4(c) following Milsum [5], we represent the mathematical equations for the relationships of storage to flow in the language of block diagrams. The mathematical terms, as they follow each other or converge, are like storage and flow similarly connected with blocks and lines.

Figure 9–4 Diagrams of one system in four languages: the symbiotic relationship of photosynthetic production and total system respiration in the biosphere (or other closed system). (a) Energy network diagram; (b) mineral cycle diagram; (c) differential equations diagramed for operational analog simulation; (d) equivalent circuit (passive electrical analog language) (see also Fig. 1–6).

In the operational analog procedure of electrical simulation, an item of hardware is arranged for each item in the block diagram of equations (such as Figure 9–4(c)). Wires represent the lines, and there are hardware modules for each of the symbols (summer, integrator, multiplier). Olsen [11] used these analog computer procedures to study transients of annual leaf fall and radioactivity passing through forests. The voltages as they rise and fall in different places in the circuit represent the mathematical terms there as in the example in Figure 9–4(c). Self-regulating properties are represented by the feedback diagram often used to explain regulators in engineering devices. On this diagram the feedbacks of mathematical quantities to a summing point are of opposite signs, one negative and one positive. See the simple example in Figure 11–1.

Compare Figure 9–4(c) with the equivalent passive circuit in Figure 9–4(d) in which the circular flow of electrons simulates minerals. The equivalent circuit (Figure 9–4(d)) has a loopback to form a closed continuous flow. Next compare Figure 9–4(d) with the same system in the energy network diagram (Figure 9–4(a)). The loopback of the mineral cycle is represented as the work done by the downstream unit on the input energy flow of the upstream unit. In the language of the energy network diagram the loopback is multiplicative, as discussed in Chapter 6. Thus the same phenomenon is manifest as a negative feedback in mathematical language and in the operational analog language; as a continuous loop without inputs in the equivalent circuit language and mineral circuit diagrams; and as a multiplicative positive feedback in energy network terms.

The circuit in Figure 9–4(c) was simulated for comparison with real data by Odum, Lugo, and Burns [10]. As automatically recorded Figure 9–5 shows the pattern of transients that followed an input of square waves representing on and off light (day and night). The horizontal axis represents over two days although the electrical output covered this in a fraction of a second. Notice the rounded rise and fall of organic matter and out of phase with it the fall and rise of inorganic raw materials for photosynthesis. These patterns resemble the real results of many microcosm studies made in recent years [2, 9, 1].

This examination in different languages of the same simple example of a two-compartment system in circular arrangement illustrates the means for translating from the energy flow diagrams to whatever other language is needed. Thus, if the compartments of an energy flow diagram are identified, we can equate them with their equivalents in the operational analog system and immediately translate the diagram to the circuits. Or

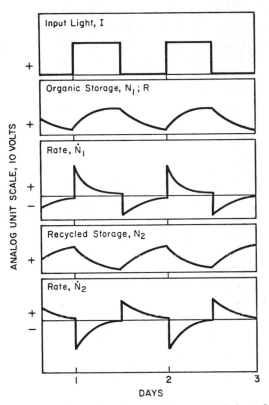

Figure 9–5 Electrical recording of analog simulation of a balanced ecosystem from Odum, Lugo, and Burns [10] using circuit given in Figure 9–4(c).

we may judge the applicability of network laws stated in other languages.

THE BALANCED ECOSYSTEM AS A DIGITAL PROGRAM

As most readers know, most kinds of calculations are easily done on the versatile digital computers once the separate steps of computation can be listed in simple English and in mathematical instructions. Then these statements are translated into a special shorthand of words that the machine recognizes. The translation is like changing English into Spanish which mainly requires someone to know the equivalent words. The list of statements is the program, often as a stack of machine cards, one statement to the card.

To simulate a model system such as the balanced ecosystem diagramed in Figure 9–4, the formula for each change in the model is listed in the sequence it occurs in the model, and the machine is instructed to repeat the sequence of calculations over and over. The machine changes its memory registers each time the calculations are made. The memory values are printed out after each round of calculations, which is regarded as the passage of one unit of time (such as an hour). By printing out the state of each property of the system and the time on each passage of the calculation loop, the record traces the consequences of the model with time.

The digital program for the balanced ecosystem model is given in English in a flow chart as Figure 9–6 using the same symbols already designated in Figure 9–4(a). In the second statement the quantities to be followed or manipulated are identified by letter in an initial input list as memory locations. In the third statement initial values of storage quantities (N_1, N_2) and coefficients (k's) are introduced to occupy the memory locations. Next provision is made to print these items on the paper for the record. Then in the seventh statement comes the first manipulation of the model. The memory location for photosynthesis per unit time (P) is changed to include the result of the multiplicative reaction of a set quantity of plants (k_2), quantity of raw materials (N_2), and light (I). What the "Let $P =$" statement says is "Erase what was previously in the memory for P and substitute what is in this statement."

Respiration is similarly calculated and registered in statement 8. The effect of P and R on the storages N_1 and N_2 are calculated by addition and subtraction. The quantities $\dot{N}_1$ and $\dot{N}_2$ are calculated by addition and subtraction and are shifted to other memory locations (X and Y) to save them for printing out, which occurs at the end of the computation round. Time (T) is increased by 1 and the new value registered. After printing out the numbers at that time, the program instructs the machine to go around again. The light (I) is turned on after 12 cycles without it and turned off after 12 cycles with light, thus making an input program of square wave. Each cycle can be regarded as an hour and each light period as a day.

The digital program does the same operation as the other simulation procedures. The digital statements followed from the energy diagram since each energy diagram module has its performance rules (such as multiplier, sum, etc.; see Appendix). Programming the balanced ecosystem model illustrates the rather automatic translation of energy circuit language to simulation languages once the properties of the model are stated in "energese."

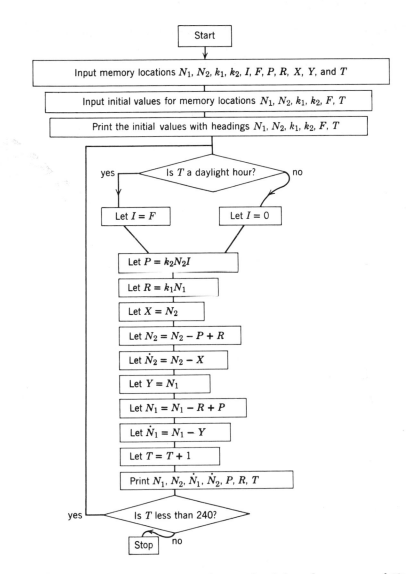

Figure 9–6 Flow chart for 10 day simulation of a balanced ecosystem of Figure 9–4(a) on digital computer. $\dot{N}$ means rate of change of N. T is taken as a daylight hour when T/12 rounded to a whole number is an even number.

METHODOLOGY OF SIMULATING A SYSTEM

The first step in considering any problem of man and nature is to identify the principal energy storage compartments, the principal energy flow pathways, and the presence of the more complex work functions such as the multiplicative action of one flow permitting a second one, or the loopback of energy downstream doing work upstream to permit flow there. Then we identify the energy sources and outflows (energy sinks). The network diagram is drawn. Quantitative numbers should be placed on the network properties and on the existing average flows and storages. Often the answer to the problem will already have been found. This process shows the relative importance of power and control circuits, the pathways of causal forces, the stability against interruptions, and the reasons for observed structure. If the details of flow must be predicted, we next set up an electrical analog or digital simulation program which can be manipulated to study the responses to varying inflows. Then we compare the simulated flows with temporal records revealed by the experimental manipulation of inflows.

AN ECOSYSTEM AS ITS OWN COMPUTER

As we have seen in previous chapters, the many compartments and circuits that constitute systems of man and nature are themselves special-purpose computers. When manipulated they provide calculations of the consequences of their own activity. The traditional argument between theoretical scientist and experimentalist concerns the best computer to use to study phenomena: the computer that the system is for itself or models such as analogous electric networks and projected patterns in our neural brain. The real system may be the most realistic calculator if it is not too slow, cumbersome, expensive, and difficult to program for study purposes. A man's brain, however, often fails to understand his own conceptual models well enough to ensure calculations that are free of hidden assumptions, errors, and unclear presumptions. The model systems force the clarity that goes with building up a whole from parts and precise function from structural extrapolations. Particularly for the great ecological and social systems, the real system is beyond experimentation and the brain systems have not been able to assimilate the complexity. Thus the model networks hold great promise.

SUCCESSIONAL PATTERNS

Whereas the separate species which are included in newly established ecological systems depend on the detail of individual adaptations, the

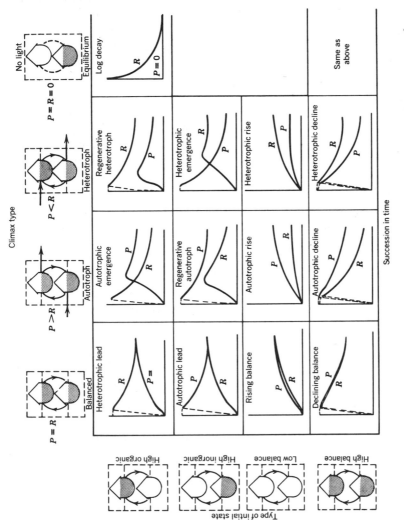

Figure 9-7 Patterns of photosynthesis (P) and system respiration (R) in the succession from different initial conditions to different classes of steady states, the latter determined by the nature of the input-outputs of inorganic nutrients and organic fuels [8].

271

overall pattern of photosynthesis, respiration, and the storages of organic matter and inorganic nutrients may be predicted from the P and R model. In Figure 9–7 are four classes of initial conditions (high and low organic matter at the start combined with high and low inorganic nutrients at the start). Across the top are steady-state conditions with different combinations of input and outflow as already discussed in Figure 1–1. There is the closed-to-matter class, the inflow of inorganics and outflows of organics (forestry and agriculture); the inflow of organics and outflow of inorganics (sewage plant or decaying litter); and the case where energy is cut off after the initial start and the ecosystem decays to the death (true equilibrium). An example is an isolated decaying log.

Systems starting with initially rich nutrients like fertilized fish ponds and fertilized fields have declining photosynthesis to the steady-state level. Systems starting with initially high organic matter like the new pond on a former forest or the newly cut forest have their respiration declining to the steady-state level. Those starting low build up in the way often represented in elementary ecological texts as the general situation.

All these patterns and others show up in the computer display scope, depending on the coefficients of photosynthesis and respiration used. The means by which one-half of the biological cycle controls the other (P and R) and passes pulses and transients gives us insights into our larger biosphere and the opportunities for simulation in prediction. Simulation of the flow of dollars of the national economy with similar models that also include the energy flows have promise for macroeconomics and national planning.

All of the models of this book may be simulated by one of the procedures indicated in this chapter, if one has data on the energy flows on at least one occasion from which to calculate transfer coefficients. For example, simulations have been done on diagrams in Figs. 1–5, 1–7, 2–6, 3–14, 5–1, 5–8($a–c$), 6–2, 6–8 and others not in the book, but most of the interesting ones involving man, economics, and war are yet to be done. We need more hands at this work. Only after simulation are we sure that the networks diagrammed behave as we think from our visual examinations or our mathematical calculations. The simulations check the consistency of our thinking and the relevance of our assumptions about what is important in the real systems.

REFERENCES

[1] Armstrong, N. E. and H. T. Odum, "Photoelectric Ecosystem," *Science*, **143**, 256–258 (1964).

[2] Beyers, R. J., "The Metabolism of Twelve Aquatic Laboratory Microecosystems," *Ecol. Monogr.*, 33, 281–306 (1963).

[3] Karplus, W. J. and J. R. Adler, "Atmospheric Turbulent Diffusion from Infinite Line Sources: An Electric Analog Solution," *J. Meterol.*, 13, 583–586 (1956).

[4] Leopold, Luna B. and K. S. Davis, *Water,* Time Inc., New York, 1966.

[5] Milsum, J. H., *Biological Control Systems Analysis,* McGraw-Hill, New York, 1966.

[6] Monteith, J. L., "Gas Exchange in Plant Communities," in *Environmental Control of Plant Growth,* L. T. Evans (ed.), Academic Press, New York, 1963, pp. 95–112.

[7] Odum, H. T., "Ecological Potential and Analogue Circuits for the Ecosystem," *Ameri. Sci.*, 48, 1–8 (1960).

[8] Odum, H. T., "The Element Ratio Method for Predicting Biogeochemical Movements from Metabolic Measurements in Ecosystems." *Transport of Radionuclides in Fresh Water Systems,* U.S. Atomic Energy Commission Division of Technical Information TID 7664, 406, 1963, 209–230.

[9] Odum, H. T., R. J. Beyers, and N. E. Armstrong, "Consequences of Small Storage Capacity in Nonnoplankton Pertinent to Measurement of Primary Production in Tropical Waters," *J. Mar. Res.*, 21, 191–198 (1963).

[10] Odum, H. T., A. Lugo, and L. Burns, "Metabolism of Forest-Floor Microcosms," Chap. I–3 in *A Tropical Rainforest,* U.S. Atomic Energy Commission Division of Technical Information, Oak Ridge, Tenn., 1970.

[11] Olson, J. S., "Analog Computer Models for Movement of Nuclides through Ecosystems," in *Radioecology, Proc. First National Symposium,* V. Shultz and A. W. Klement, Jr. (eds.), Reinhold Publishing Co., New York, 1963.

[12] Olson, J. S., "Equations for Cesium Transfer in *Lirodendron* Forest," *Health Physics,* 11 (12), 1385–1394 (1965).

[13] Wyckoff, R. D. and D. W. Reed, "Electrical Conduction Models for the Solution of Water Seepage Problems," *Physics,* 6, 395–401 (1935).

10

PARTNERSHIP WITH NATURE

When the energy budget of man and that energy under his control was but a tiny percentage of the energetic system in which he lived, he simply was not a manager of nature; his survival was provided for him as long as he made a loop-reinforcing circuit, doing work on a part of the system and thereby encouraging that part to supply him with food and clothing. With new, rich energy flows man can control and manage enormous quantities of the world's energy. Increasingly he is affecting the whole biosphere. Man's cultural inputs and outputs begin to approach those that maintain the stable composition patterns of air and ocean. The rising concentrations of carbon dioxide and atmospheric pollutants are examples. As man's role increases, the productive buffer of the natural system becomes less and less capable of offering protection. As man's energy flow increases he has become more and more influential on the design of ecosystems, and it becomes more and more necessary that he provide a reasonable system coupling nature with his culture (Fig. 10–1). With a phrase in *Governing Nature*, Murphy [7] traces "the changing currents of public opinion and legal practices which have moved away from the assumption of nature as a free good." The management of nature is ecological engineering, an endeavor with singular aspects supplementary to those of traditional engineering. A partnership with nature is a better phrase.

THE NETWORK NIGHTMARE

The difficulties of managing nature and man can be stated in circuit terms. Let us first fantasy the nightmare of an electronics technician. After a week of exhausting tedium, soldering circuits and completing a

274

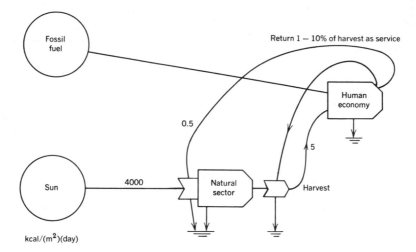

Figure 10–1 Ecological control engineering defined and illustrated by loopbacks of service of less than 10 percent of man's harvest from nature.

large network of wires connecting thousands of tubes, transformers, and transistors, he goes to bed with the feeling of a design completed. Then with the veil of the dream the parts begin to breathe. Next he sees them grow and divide, making new parts. Then the wires become invisible. The new and old parts disconnect themselves and move into new patterns, reconnecting their inputs and outputs, replacing worn members, and together generating functions and forms not known before. What was neat and known becomes unknown. Soon the new system with its vast capabilities is growing, self-producing, and self-sustaining, drawing all the available electric power. Our hero awakens when he pulls the switch removing the energy source. To some visionary engineers the nightmare may seem a preview of a machine world. Our ecosystem, however, is already the nightmare.

Already covering the planet earth are billions of breathing parts, each with a program of functions. These parts also divide, replace each other as they wear, and connect themselves to inputs and outputs. Like the flexible circuits in the dream, they develop immense complexities that drain available power sources. These parts are the living populations, and the circuits are the routes of flow of materials and energy.

The men trying to understand the patterns, to predict and design new combinations and harness the great systems of man and nature, experience another nightmare, one of uncertainty. Like the electronics technician in his dream, the student of the environmental system is given no circuit diagram and the pathways are almost invisible. These systems of fantastic complexity include man, and his survival depends on their function. If such systems change, so may the human roles therein.

STEADY STATES OF PLANETARY CYCLES

With the simulation experience of Chapter 9 in mind (Fig. 9–4) the reader can visualize the behavior of a water tank which has a storage of

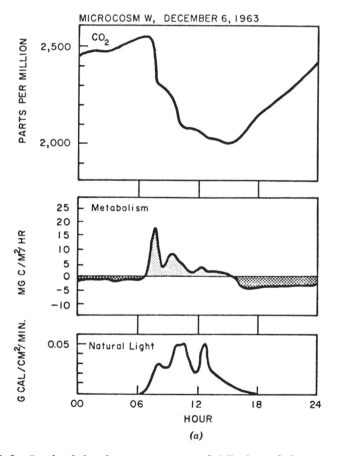

Figure 10–2   Graph of the characteristic rise and fall of metabolic quantities with the alternating daytime photosynthesis and nighttime consumption. (*a*) Small terrestrial microcosms (terraria) (from Ref. 16); (*b*) small aquatic microcosm (Armstrong and Odum) [2] with blue-green algal mat (balanced aquaria); (*c*) reef of oysters and plankton in a circulating tank [13].

water that depends on the balance of inflows and outflows. The outflows depend on the pressure of the amount of water. Since the outflow increases with storage, the tank tends to fill to a level that is steady for a particular inflow rate. Many of the living and nonliving storages of the biosphere follow this pattern which is also simulated by electrical storage units (capacitors). Such storages are said to be density-dependent and because of the negative property of the outflow are said to have negative feedback stabilization. (They have negative feedback loops when their equations are diagramed. See Fig. 11–1.).

The size of a storage can be readily changed with different coefficients of the outflow. For example, the water tank level would fall to a lower steady state if one opened the outflow valve to a larger dimension. In

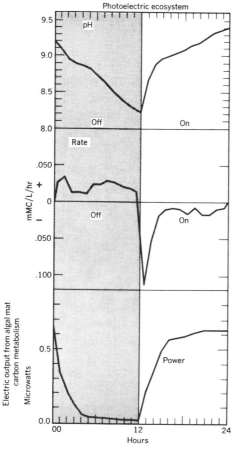

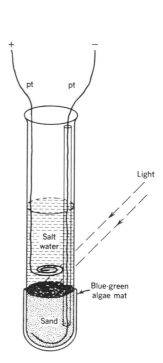

(b)

the biosphere the organisms are the control valves on the rates of flow of the minerals and gases from storages. If evolution changes the populations of organism, the outflow rates of the earth's cycles are changed and the amounts in storages change. For atmospheric carbon-dioxide this principle was demonstrated in terrestrial microcosms [15] (see Figs. 1–6 and 10–2). Each microcosm had slightly different biota and the levels of carbon-dioxide which develop in different systems ranged from 300 to 2000 rpm. If the present 340 ppm of the biosphere were increased to

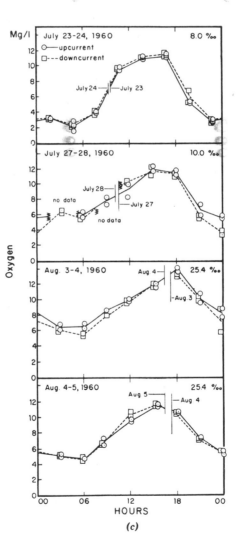

(c)

2000 by a change in the ratios of plants to consumers, the greenhouse effect of carbon dioxide in preventing outward nighttime heat radiation would raise the earth's temperature, presumably melting the ice caps and flooding the coastal plains of the world. This much-discussed possibility is only one of the changes in flow that may be expected if further power injections change the world ecosystem's composition. Now, increasing turbidity in the air from pollution is reflecting light in many areas, lowering temperatures and agricultural production. Somehow we have to prevent major disturbance of the coefficients of the mineral cycles.[1]

ECOLOGICAL ENGINEERING OF NEW SYSTEMS

The millions of species of plants, animals, and microorganisms are the functional units of the existing network of nature, but the exciting possibilities for great future progress lie in manipulating natural systems into entirely new designs for the good of man and nature. The inventory of the species of the earth is really an immense bin of parts available to the ecological engineer. A species evolved to play one role may be used for a different purpose in a different kind of network as long as its maintenance flows are satisfied. The design of manmade ecological networks is still in its infancy, and the properties of the species pertinent to network design, such as storage capacity, conductivity, and time lag in reproduction, have not yet been tabulated. Because organisms may self-design their relationships once an approximately workable seeding has been made, ecological network design is already possible even before all the principles are all known.

We are accustomed to species occurring together in particular associations, but when conditions change, species may assort quite differently to effect a workable metabolism. For example, when we construct saltwater ponds, we find freshwater and marine organisms together in new combinations, rotifers with barnacles and waterbugs with marine algae. When we walk along the shores of the Salton Sea in California where combinations of marine organisms from various places have become established in a new kind of network for a new environment, we tread

---

[1] The equation for asymptotic exponential growth is sometimes called Von Bertalanffy growth and is characteristic of any kind of charge-up process that balances a steady input with storage controlled output. $Q = RCJ \ [1 - e \ (t/RC)]$ where Q is the quantity in the tank, J is the inflow rate, RC is the time constant, and t is the time (see Ref. 22). (For application to forest ecosystems see Ref. 17).

on the shell heaps of the blue barnacle instead of on the shells of the more familiar mollusks of the present seas.

We are reminded of the strange and different associations in the fossil beds of past ages. We are accustomed to thinking of these differences as the consequence of other organisms coming into existence through the evolution of species. Yet equally contrasting associations now found in isolated basins of the arid regions of the world indicate that varying networks are self-designed with changing circumstances. Different species were in existence, but the variation in the dominants of past seas may have been as much related to conditions eliciting adaptive network designs. The permutations and flexibility available for future unique and new ecosystem designs may be equally as great. The strange networks on islands and in isolated lakes are samples of what can be accomplished with ecological engineering.

Fortunately, the parts from nature's bin are self-switching and adaptive; groups of species acting together rapidly develop interconnecting networks of food flow or are eliminated. Thus, multiple and continual seeding of many species into a particular environment will aid and accelerate the development of a stable network, most of the work being done by the self-designing tendencies of any association of species.

MULTIPLE SEEDING AND INVASIONS

One of the means for developing stable new ecological designs for new environments is multiple seeding; many species are added to the new ambience while conditions are maintained as they are likely to continue. In microcosm studies in which this kind of work has been done, the species go through self-selection of loops, producing a stable metabolism and a complex network within a few weeks. It is very tempting to recommend this method for creating new designs in larger-scale systems such as lakes, rivers, and lands. About transplanting species, however, there is much controversy. From some bad experiences with invaders in the past, considerable fear exists that such disruptions of nature lead to further upsets of previously stable environments. Species removed from their roles in a stabilized system no longer find the cues by which their own behavior adapted them to the well-being of their system and hence of themselves. They may find new energy sources and at first reproduce in great numbers.

Species removed from their normal loops may have no controls at first other than those provided by man. If they grow too fast, draining some aspect of the host system, they will eventually achieve their own control because their circuit will not be loop-rewarding. A new parasite lowers the output of its host until both are reduced in role.

When an existing system is disturbed and new species are added, there may be great fluctuations and pulses during the self-designing process. Some of the invasions of exotic organisms such as the gypsy moth and chestnut blight have been extremely detrimental. Other invasions and transplantations have enriched the receiving systems with fast-growing trees, and birds that are able to live in cities. An invader may be filling a void in a new situation created by man, although men are tempted to blame the invader for causing the disruptions. In our world of disturbed nature we may need these invasions. Sometimes a rare native species will emerge to serve the new situation.

When man has already disturbed and displaced the natural system, conditions are different and new species may be required to form a new network. The rapid success of some invaders is possible partly because they are able to use portions of the disturbed system that have fallen into disuse.[2] Man's agricultural efforts are really simple ecosystems with all kinds of open potential energy kept isolated by the insecticides and other stringent management measures. The policies of the federal agricultural agencies against importing species are one means for protecting the simple agricultural networks. Unfavorable experiences with invaders displacing agricultural species have conditioned many scientists to oppose transplanting as a conservative play-safe policy.

Perhaps conservatism was justified as long as natural systems were little disturbed and knowledge was lacking. Now, however, the natural systems are mostly disrupted and man is integrating small separate systems into one larger system anyway. Perhaps now multiple transplantation is needed to develop new designs. There are likely to be temporary imbalances, epidemic-like population growths, and depletions from overgrazing, predation, and parasitism during the adaptation of a new design, but the final product may very well be a worthwhile new pattern of stability. This kind of self-design will take place anyway, but man can help by providing multiple-seeding experiments in selected situations.

Systems may be developed to regenerate heavily polluted waters. For example, the development of plant varieties tolerant to toxic heavy metals such as lead and copper is described in Ref. 4. Resistance to DDT by houseflies is known [9].

---

[2] Charles Elton's *Ecology of Invasions* [6] documents the many famous cases of species from other areas having large success and becoming principal members of new areas.

THE IMPLEMENTATION OF A PULSE

It is a principle that complex systems develop great stability with few spurts of food and fruit production when there are no pulses in their climates. Tree growth is slow and steady without rings; flowering and fruiting are gradual and small at any one time. Total photosynthesis is great, but the system is composed of species adapted to steady, small, and diversified flows. It is a difficult system to harness for large power flows. Man in such systems was like the Pygmy in the Congo and the deep-forest Indian of the Amazon.

If, however, we can use available energies to provide a sharp seasonal pulse, the system of nature redesigns itself and is reduced in diversity because energies are stored to last over the pulse periods and are diverted into other special adaptations required by the pulse. By creating a pulse, perhaps by controlling the water table or by covering plants with black plastic and introducing new plantings to accelerate the self-design, a simpler system with a large net yield may replace the complex one with many low yields.

ENERGY CHANNELING BY THE ADDITION OF AN EXTREME

The great diversity of plants produced by the division of labor in natural systems of the stable moist tropics is especially difficult for man to harvest. In effect, when there is stability the energies of the sun produce a complexity that exceeds anything man can accomplish with the same energy budget. In the rain forests and on the coral atoll and the oceanic islands in even climes, man must join the system and be content with a minor role at that.

If, however, it is possible to add an extreme to which the natural components must adapt, diversity drops out and the yields, although less, may be sufficiently channeled to provide a harvest for man. For example, if a coastal area has suitable topography, it may manage its tidal flows so that a high briny salinity of about 4 to 5 percent is maintained. The consequence is an abbreviated community of algae, plankton, and a few species of fish with energies channeled into high production yields. The beautiful coral reefs provide wealth where biochemical diversity, recreation, and stability are needed; the briny basis supplies protein and a food chain for man.

Where there is a sharp season, animals may migrate. For example, wet seasons see many animals dispersed over savannas and marshes, but a dry season sees a contraction and a natural harvest system bringing a concentration of animals to the water holes. Systems with sharp energy seasons may see shrimp migrations from bays to estuaries and rivers. If a food pulse is introduced in the water management of an area, and if stocks with migratory tendencies are added, there is a selective survival of the migratory populations which establish movements in phase with the food availability.

MICROBIAL DIVERSIFICATION OPERATORS

Tropical grassland animals, marsh animals, and to a lesser extent forest animals make use of the microbiological complexes to change the chemical diversity of the plant material into an organic soup usable in their own meat production. E. P. Odum [12] suggests a detritus agriculture following the pattern of many natural systems. Why not set up composting and natural silage systems and induce the chemical changes on an organized scale? Can we duplicate the microbial systems in the stomachs of ruminants? Can we duplicate the microbial complexes in the litter of the forest that support the fairly large stocks of earthworms? Perhaps the extremely high photosynthetic rates of early successional species in the moist tropics can be utilized as an input to a microbial vat system, the output of which will be proteinaceous animal products.

One established system already extensively used is the discharge of leafy materials from the land into canals and ponds where fermentation and decomposition lead to very large fish populations. Examples are the canals among the sugar cane fields of Guayana. Apparently the leaves have sufficiently high protein content and the aquatic conditions are adequate to develop high rates of cellulose decomposition. This is one system which, if it were better documented, might be considered a sure method for protein concentration to accompany a terrestrial basic food system.

ECOLOGICAL ENGINEERING THROUGH CONTROL SPECIES

Where man has only a relatively small budget of potential energy available for use in a large area, it is possible that given enough understanding of the workings of the system he can apply his energies to control its larger organisms. They, in turn, operating through the food chains, can exert control over vast areas so that those special outputs within the system's energy budget will be focused to man as he may direct. Managers of complex factories with fingertip exertions manipulate the outputs because they know how the controls work. Presumably this amplification of effort is equally possible for man at the head of natural systems of forests and oceans. Should the auxiliary fossil-fuel and nuclear energy sources fail, this control is one of the bright prospects for man, but he must obtain the knowledge now while he has the energy surplus available to pay for research and experimentation.

When potential energies are in great excess, man may remove the preceding system and set up something new of concrete and steel. When as in his prehistorical beginnings he has no energy other than that provided for him from a tightly organized natural system, he has little left over with which to manipulate. In the intermediate situation, with some fuel supplements but not enough to dwarf the environment system completely, he must practice ecological control engineering. In the terms of the network diagrams, man needs to contribute about 10 percent of a system's energy upstream in control work in order to stimulate the system toward a chosen yield (see Fig. 10–1).

THE CROSS-CONTINENT TRANSPLANT PRINCIPLE

In the tropical forest the balance of nature apparently provides just enough herbivores and carnivores to keep the levels of each species regulated so that not too many of any one are allowed to develop. Consumers and diseases help serve the constructive group role of maintaining stability and well-balanced numbers among the many species. Epidemic actions on single species stabilize the whole. Any attempt to form plantations with a native species is likely to set in action the mechanisms that had evolved earlier to keep all species regulated at safe low levels. Past efforts to produce Brazilian rubber on plantations in Brazil brought on epidemic destruction and financial disaster.

When species are completely removed from the continent where they were part of a stable system, growth and production often have a wholly different order of magnitude. Apparently the cues from the environment that provide limits, as well as diseases and specific control insects, are absent. The transplanted species are often the best for net yields. Australian trees have grown especially well away from Australia, and plantations of Brazilian rubber have done better in Africa and Asia.

For new net-yielding varieties we should explore further the principle of intercontinental transplantation. When and if the plantations of imported species are invaded by transplantations of disease or insects, with present knowledge we have to inject chemical control know-how and the aids of a fossil-rich culture or import natural controls.

MAN AND THE COMPLEX CLOSED SYSTEMS FOR SPACE

As man begins to be important in his biosphere, he must consider what system is sufficiently compatible with his inputs and outputs to maintain a viable atmosphere, an adequate water cycle, and other favorable conditions. The design problems are the same as those for the minimal system of man and nature capable of export to the moon for support of human life in space.

Small closed systems have been studied in the classrooms for years, but now new efforts have been made to develop a closed system sufficiently large in size to include men and the diversified food network capable of supporting him. Stimulated by the space program, considerable effort has been expended on the study of closed systems involving one or two species, but no stable support of man has been achieved. We really cannot generate from one or two species the complex input required by man. We cannot provide a complex phone switchboard with an input cable of only two or three wires. Not only must food be made and waste regenerated, but long chains of work activities are required for arranging molecules into the complex nutrition and energy concentration required for man. Having evolved in nature and requiring stabilization in a complex system, man and his needs cannot be simple. The carrying capacity of an area is not figured only on the basis of the energy necessary for producing man's calorie input. Also required are energies for regulating, stabilizing, and diversifying the network of minimum complexity. Areas for one man given in Table 4–2 range from 0.001 to 640 acres depending on fossil-fuel supplements.

Hence something of a controversy has existed about what area of the sun's energy is required to support man, whether on earth or in space. How much of the old systems must be retained for stabilization and regulation? Or what part of the old systems must be designed into the new ones? The extremes of the opinions on man's needs in a closed system range from a square meter to acres of solar energy.

Evidence that the way to stability is in complex multiple-species systems comes from studies of small closed ecological systems which balance the net photosynthesis of the day with the total consumption during the night. The measurements of oxygen and carbon dioxide show that such systems (Fig. 1–6) reestablish gaseous balance in 48 hours and have a steady undulation of atmospheres in a 24-hour rhythm, even when subjected to such disruptions as gamma irradiation (short x-rays) of 25,000 roentgens, a dosage twenty times that necessary to kill people (Fig.

10–2(*a*)). The dependence of photosynthesis on respiration and vice versa are so well coupled in some aquatic systems that changes in temperature have little effect (see Fig. 10–3, Ref. 3). The increase in respiratory processes at higher temperature is braked by the photosynthetic rate of providing needs and using up products. The regulating mechanisms and the electrical analog of this cycle were given in Figure 9–4.

The much needed but expensive experiment is diagramed in Figure 10–4. Man is kept in an armory-sized structure into which all the materials of the biosphere are introduced along all kinds of biological components, especially microbiological ones. By the process of learning, loop selection, succession, and evolution discussed in Chapter 5, a system of man compatible with his food and waste productions in a restricted environment is likely to emerge. The same people might not remain inside for the full period, but several years of continuous occupancy would probably be required to begin to develop a compatible system

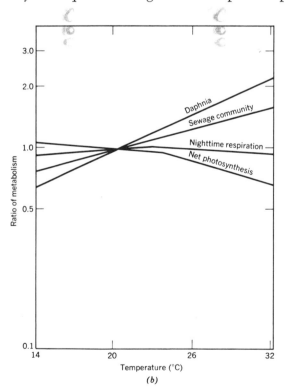

Figure 10–3   Graph of the response of balanced aquaria to changes in temperature [3].

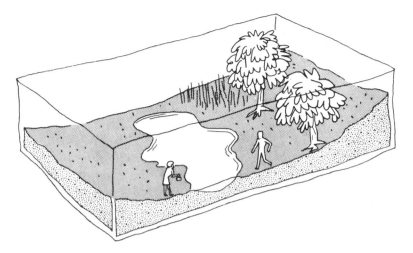

Figure 10–4 Adaptation of man to nature in accelerated self-design in a terrestrial capsule experiment following multiple seeding of an armory-sized closed system.

with the stability already proven in even smaller microcosms. Judging by the minimal land areas that have supported man without fossil fuels in the past, one or more acres are required to give the energy concentration, the diversity of nutrition, and the stability of networks.

An important limit to man's invasion of space is his life support, his need for a regenerative, complementary ecosystem. Something of a national fiasco has developed, limiting the ability of the United States to put man in space for any length of time. In developing biological regenerative systems for space, the National Aeronautics and Space Administration from 1960 through 1969 refused to recognize that multiple-species designs are required for stability and that this energy cost is unavoidable. The bias of its supervisors was for hardware control at even greater energy costs, bypassing the miniaturized microecosystems. They considered only cultures of a single population which are always unstable except when under heavy and costly chemical controls. The designers argued that the multiple species system was too inefficient in requiring too much space. However, the efficiency argument could be made for all the hardware that had to be included, fueled, and ejected from the earth's gravity with the moon trips. The ecological formula would not require great weight, dual maintenance back-ups, or much fuel. It would require some kind of expandable surface. A large terrestrial capsule was never given a trial in spite of NASA's great expenditures. For more on this argument see Ref. 5, 15, and 22.

There is a similar failure to understand that man is dependent on his biosphere's naturally diverse system for regeneration of life support. The world poses the same problem as the space capsule. Management of nature and ourselves in this instance is the same thing.

COMPATIBLE LIVING WITH FOSSIL FUEL

While man is running on the great excesses of fossil fuels, he is creating a huge new system with input and outputs that never existed before. New outputs are insecticides, detergents, pulp wastes, heavy toxic metal flows, waters loaded with heavy organic loads, and wastes with radioactive isotopes. New inputs are his diggings into the various fossil storages, his massive disruption of the land surface, his short-stopping of the water cycle, his takeover of the better photosynthetic surfaces, and so forth.

Nothing about man's present system is balanced, for his inputs come from geological storage or from the energies that used to go to balanced systems. His wastes are new in quality so that a system of organisms capable of using them and complementing man has had little time to evolve. Possibly organisms able to redesign a workable combination and deal with the new conditions exist somewhere on earth, but they have had no time to go through the selection and circuit-reinforcing process necessary to develop a new kind of ecological system to match the new outputs and inputs of man.

Figure 10–5 shows a large-scale experiment which should be carried out to help develop compatible systems that will loop back to man the inputs he needs from the environment, with reward stimulation to both man and his inputs to make these loops competitive over others. The area chosen should be seeded with microbes, plankton, fishes, and terrestrial animals from all over the world. The representative inputs of man's ecosystems should also be added. This ecological engineering project might then through the self-designing process develop a compatible system.

Following its use in microcosms, a small-scale test of this principle was carried out in North Carolina marshes under support from the Sea Grant Division of the National Science Foundation in 1969 to 1970. City-treated sewages were mixed with estuarine waters and run into three ponds continuously, the ponds being well seeded with estuarine organisms. A regime of fluctuating algal blooms developed and with it some members of the natural system with adaptations to varying oxygen to develop a food chain with considerable diversity with substantial numbers of blue

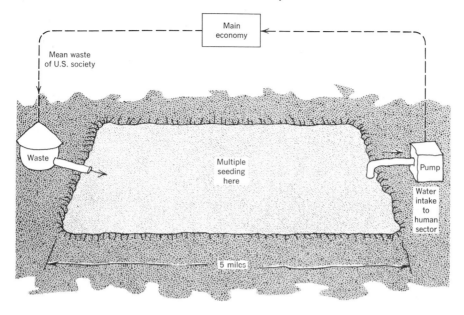

Figure 10–5 Principle of adaptation of a regenerating system to a waste flow characteristic of modern civilization.

crabs at the top of the food chain. Some wastes of the city of Morehead City, N.C. were being mineralized by an ecosystem that was processing "tertiary treatment" free. Control ponds receiving tap water and estuarine water had an ordinary kind of estuarine community. The pond experiments on a small scale suggest that the self-design principles are correct, a useful general procedure for blending the human sector with the non-human sector.

HOW TO PAY THE NATURAL NETWORKS

If man withdraws from a natural network a harvest of fish, quail, timber, or other beneficial items for his use, he puts a drain on that system. Whereas the circuit he is draining was formerly competitive, it may now be displaced as a principal flow by some other species because of the extra demand placed on the flow. If man withdraws materials and energy without paying for them in a receivable currency, he has done the opposite of rewarding loops; he has broken loops and punished the loop that he finds useful.

If, however, man pays nature in a feedback reward to that loop and the reward has as much energy-benefiting effect as he has withdrawn, and especially if he closes the loops involving nature and himself with limit-breaking flows and matching chemical ratios and rates, he will ensure the survival of the product of his interest. As indicated in Chapter 4, systems seem workable when the loopback reinforcement is only a small percentage. The cheapest way would be to use man's waste flows to stimulate the circuits of his interest. Yet there are relatively few systems of nature that will find man's present wastes a stimulus and in turn be useful to man. As fast as such systems are found, they should be labeled and transplanted. An example is the *Spartina* estuarine marsh in Calico Creek, Morehead City, N.C. where the grass, snails, and above water ecosystems are more fertile from release of treated sewages than comparable marshes without it.

Empirically, wildlife management has worked toward some of these goals. Although hunting takes its toll of quail, energy can be and has been put back into the environment to guide the plantings toward a system favorable for quail. If a given food chain for a commercial species is characterized by particular chemical ratios, such as a nitrogen-to-phosphorus ratio, why not then treat sewage waste to produce these ratios so that the waste outflow will simulate the food chain?

A device to ensure that nature is paid for its work might be based on the dollar-energy relationships given in Chapter 6. In using a natural

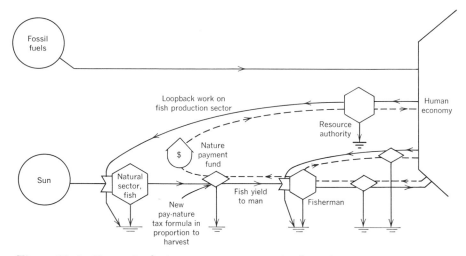

Figure 10–6 Economic device to arrange proportional service payments to nature for its service to man in order to maintain loop reinforcement of man and his yielding systems.

resource an individual or industry could be required to contribute money to a governmental fund in proportion to the natural energy budget of the goods received from nature. These funds would then be used to stimulate the natural system, with efforts directed at the removal of limiting factors and the reinforcement of usable resources. Whereas we do this in principle with state wildlife, automatic formulae are needed.

In exchange for fish, for example, money is contributed to a fish-stimulating fund which in turn pays other men to do work for nature. The dollar flow helps ensure a loopback of men's work to couple nature and man in reward loops. Figure 10–6 is an energy network diagram of this process. Some governmental programs that tax expenditures on hunting and fishing and put the money collected into research and stimulus of the environments are already approximating the model, although not with sufficient system-by-system detail. There is no consensus as to the loop reward fraction necessary to keep the natural sector viable.

THE CITY SEWER FEEDBACK TO FOOD PRODUCTION

As populations rise and converge in urban areas, the concentration of city wastes with their rich regenerative mineral substances becomes more important, not only in the great costs of disease prevention and safe disposal but as a potential means for completing an internal effective loop between the input agriculture of the region and the city consumers. Perhaps regionwide planning can work for a loop from the city wastes back on to the producing lands. The areal flow pattern, if constructed like other ecosystem designs, must diverge as in Figure 3–5. Such a sharp injection of fertility would do more than raise the photosynthetic potentialities; it would tend to replace the normal complexity with a few channeled producers. Here is a means for investing the wealth of the fossil-fuel culture into a loop system to increase the production of an area and to experiment with the coming principle of regional, biogeochemical organization. Better yet, if in the great diversity of tropical animals a complex can be found that will eat the wastes fairly directly after some microbiological preparation, what better means of processing the organics out upon the pastures. See Table 1–1 for element ratios.

SPECIALIZATION OF WASTE FLOWS

The high concentrations of chemical substance in wastes are actually rich reserves of energy and substrates capable of supporting special processes for special purposes, but it is difficult to treat them in recycling systems.

The organisms of nature tend to develop specialized recycling mineral circuits with their many micro-organisms, and man may need to separate specialized waste flows so that individual networks of regeneration can develop and eventually be connected back to form yield-producing loops. Wastes would therefore have to be processed in isolated flows. Specialization requires organization. Some waste problems may need regional solutions and pipe systems larger than those of single cities.

Authorization for the establishment of a new industry might hereafter require that the complete loop of the manufacturing process be specified, including provisions for the regeneration of wastes and feedback as a useful reinforcing injection into another part of man's system.

PRODUCERS FOR THE ECOSYSTEMS TASK FORCES

For many of the world's systems we do not have enough knowledge to draw the energy diagrams, predict consequences of treatments, and manage nature. Task forces are needed to study these systems, and many are at work already. See Fig. 1–4.

The principles of an energy network may provide large teams of scientists with an organizational basis for studying the great, complex ecosystems and regions of man and nature. Although much of scientific study is individual and is generated from the particular abilities of scientists, a general recognition of the network concept of synthesis allows strong personalities with hard-driving initiatives working separately to combine data about the function of whole systems. By developing a summary network diagram and quantitative data for the flows, it becomes possible to predict responses of the whole network to impending changes such as pollution, new inputs, addition of parts, fire, introduction of isotopes, fertilization, harvests, radiation destruction, and shifts in energy availabilities.

In the search for solutions of many current problems, large areas of nature and groups of men are being studied by teams using the network concepts for synthesis. For example, there was the Interocean Canal Commission Study of the feasibility of digging another Panama Canal using nuclear detonation (under Battelle Memorial Institute), the International Biological Program studying the ability of major land and water systems to produce organic matter, and several programs on the effects of radiation in pine forests, rain forests, and deserts sponsored by the Atomic Energy Commission's Division of Biology and Medicine. Effort to study a rain forest at El Verde, Puerto Rico was made by an AEC team of the Puerto Rico Nuclear Center with visiting scientists [16]. Some idea of the sequence of activities is given in Figure 10–7.

An effective procedure for a system task force might be to have waves of specialists do their jobs in a series of phases, spending approximately half their time gathering the network data they owe to the synthesis and half in exploring interesting aspects of phenomena they discover at the level of their special interest and assignment. The phase waves are as follows:

Phase 1: Description and identifications. Well-trained subprofessional systematists under the technical supervision of senior systematists at museums and elsewhere, but under the field direction of a phase supervisor, focus on a study area, and perform such tasks as tagging the plants, making reference collections of the animals, and identifying bacteria with the immediate aim of publishing documents and systematic keys that will make later phases possible.

Phase 2: Network diagram. Ecologists experienced in natural history and autecology, but supervised by a systems ecologist, next join the project. Their mission is to draw a master compartmental flow diagram showing the principal food chain routes and groups of species that have some common basis for being lumped together in group compartments. The species constituting 90 percent of each general trophic category are compartmentalized separately; the rest are put into a 10 percent miscellaneous category.

Phase 3: Chemical and energy flows. Systems ecologists perform the work of this phase, gathering the biomass values, metabolic rates, and chemical constituents of each compartment so that they may be examined by systems analysis. These data are added to the qualitative diagram of phase 2, making it quantitative. The most important value is the total metabolism and it should be measured first.

Phase 4: Experimental simulation. Now that the broad features of the system are in the form of network power flows, theoretical ecologists or engineers can study the model with computer-simulating methods to predict outcomes of treatments.

Phase 5: Management. With the model developed so that it adequately simulates observed phenomena, it can be consulted whenever recommendations must be made for managing the system to fulfill particular goals. When responses of the real system differ from the simulated results, the model is modified. Gradually as the fit improves, real prediction and savings are possible.

ENERGY-BASED VALUE DECISIONS

In many planning decisions concerning the development of industry and cities in competition with previously existing wilderness and other natural

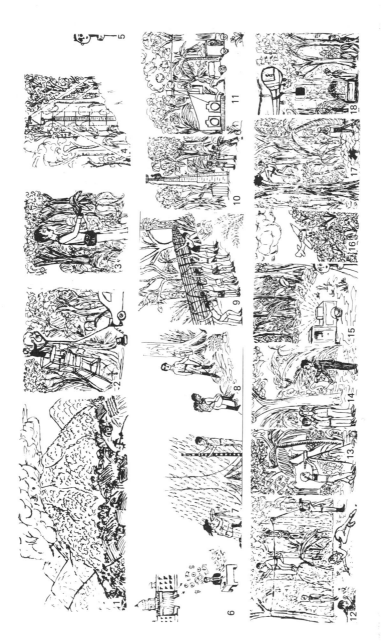

Figure 10–7 Sequence of task force activities in an Atomic Energy Commission project on the mountain rain forest in Puerto Rico in which gamma-irradiation effects were studied (drawing by Arn Odum). *1957–1961:* (1) hit-and-run studies in the Luquillo Mountains; (2) metabolism measurements in a tree house; (3) epiphytes found radioactive with fallout. *1962:* (4) cutting a biomass prism; (5) see the health physicist who appears again below; (6) AEC funds the rain forest project of the Puerto Rico Nuclear Center and orders go out; (7) survey with transit; (8) trail building, tree tagging; (9) tower construction; (10) meteorological systems and wire telemetering; (11) temporary power as generators go operational. *1964:* (12) visiting scientists swarming in the forest and down cable car; site visit; (13) U. S. Army teams map the trees and boulders; (14) AEC committee on site visit; (15) preliminary radiation test with 6 curie source; (16) barnstorming photographs; (17) accident—fall from tree; (18) in comes the

294

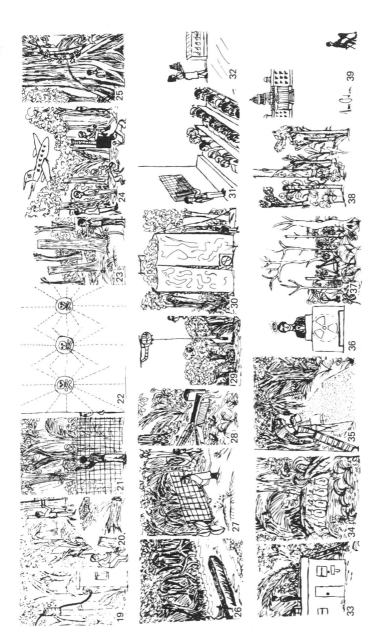

study; tree growth study with metal vernier tapes; (21) eight-foot hog fence erected around the radiation area; (22) three months of radiation; (23) pruning the cut center as a control area; (24) visiting scientists swarm back; (25) comparative studies in the virgin forest of Dominica. *1966:* (26) comparative studies in Darien, Panama; (27) spraying the ground with isotopes for study of mineral cycles; (28) microhydrological studies—flume and weir; (29) out goes the source pig; (30) giant cylinder working to measure forest breathing; (31) all-day symposium on the project at American Institute of Biological Science; (32) processing mountains of machine cards and date reduction. *1967:* (33) fallout studies accelerate; (34) covering the forest with soil percolation lysimeters; (35) a dust effect experiment; (36) symposia at the Radiation Research Society and at the Radioecological meetings; site visit by AEC committee; (37) the radiation center in successional recovery mainly up from the ground; (38) the north cut center in successional recovery resprouting from the trunks; (39) everybody writes results! Book published in 1970 [16].

295

economies, the wilderness values to society are underappraised because they have no easily recognized dollar value that can be compared with the figures available for the proposed new enterprise. Perhaps the energetic common denominators can be employed to evaluate all uses in the same terms so that planning boards can act fairly, protect the public interest, and develop patterns for the energy network that is best for man's survival.

In determining the value derived by a population of people using a wilderness area, we must consider the work done by both man and nature in arranging for the recreational and other experiences, including visual and chemical buffering. Thus, per acre, we may add the calorie cost to nature in developing the wilderness through work (its maintenance cost), the calorie cost to people of enjoying it, including travel and their individual metabolisms, and various work expenditures by society that advertise, produce, and sell commodities used in the recreation.

An example of recreation values interacting with an ecosystem is the system of marine bays near Corpus Christi, Texas used for many recreational and industrial activities. The contributions of some of the energy flows including calorie equivalents of economic activity are included in Figure 10–8, an energy diagram for the interaction of the ecosystem and human economies. The importance of the natural metabolic budget is apparent whereas it would be ignored in usual economic evaluations. Even the incomplete simple model in Figure 10–8 shows many energy flows with which value may be associated. A value web is like a food web with many connections and branches, not readily kept in mind. Little wonder that evaluating the uses of nature leads to argument and controversy.

The diagram in Figure 10–8 illustrates the dual work of nature and man in developing recreational value. Many special values such as those of historic sites and art may also involve the work done in developing appreciation in the human population. Some treasures are almost priceless because great energies were consumed in the past to create an appreciation of the site or object. For example, the large energies of wars or centuries of preservation and care may have been contributed.

REPLACEMENT VALUE OF ECOSYSTEMS

Many kinds of public actions of sweeping importance to natural systems are justified on various kinds of cost benefit ratios based on money in such a way as to ignore the important values not part of the money economy. In some public hearings and court cases in 1969 the side of the greater good has been reinforced with ecosystem network diagrams which

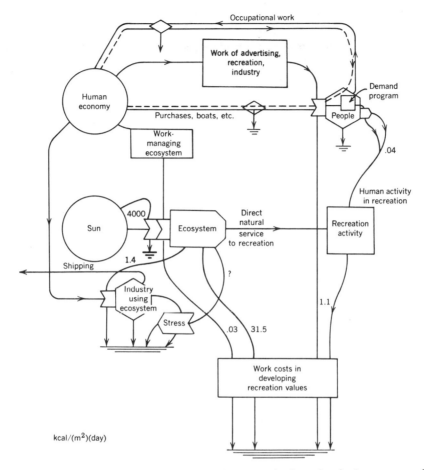

Figure 10–8   Diagram illustrating some of the work done by both nature and man's economic system in developing recreational values for man from natural ecosystems. Numbers are for the complex of marine bays near Corpus Christi, Texas with an area of 228,480 acres. Economic activity was evaluated by Anderson [1] with recreation, tourism, and fishing $166/(acre)(yr) and oil, gas, shell, cooling waters, and savings on shipping $203/(acre)(year). Using $10,000 kcal/dollars work equivalents were calculated. Energy contributions of personal activity by 13,914 visitors per day was converted using 3000 kcal/(person)(day). Management by state and federal agencies was estimated as $1 million annually.

can readily show the fallacies and incomplete calculations by developers such as the Army Engineers and large corporations. Figure 10–8, for example, shows the greatest energy values without dollar equivalents.

However, it is possible to reverse the calculation which we did on work equivalents of dollars. We can put dollar values on the ecosystem energies using the dollar/calorie equivalent of the whole economy as a rough approximation. For example, in an incident in North Carolina, surveyors cutting a swath through a public recreation park with climax forest claimed that their damage to the public per acre was $64 for the wood they cut. Public protest reflected the deep sense that this was not just.

First consider the actual value in energy units and convert. The value as a public recreation and life-support system is its replacement cost. To replace complex, diverse, and beautiful forest requires about 100 years. The photosynthesis per square meter of a forest may be approximately 40 kcal/(m^2)(day). The dollar equivalent of work driven by organic fuels is about 10,000 kcal/dollar. With 4047 m^2/acre, the dollar value of replacement of an acre of this forest is $590,000 per acre. Losing the development value of 100 years for an acre of land is a major loss. A single tree of about 100 years of age is estimated in this way to be worth $3000.

LIFE-SUPPORT VALUES OF DIVERSITY

One of the tensions between conservationists and economy-minded decision-makers concerns the importance of preserving the complexity and diversity of birds, flowers, and rarer members of the natural ecosystems. Also in argument is the disposition of research funds between study of toxicities of wastes on man directly and responses of the ecosystems to our increasing levels of wastes. To many the birds and bees have seemed to be an extra luxury to economy and survival of man. Conservationists have challenged this, but rarely found concrete language to explain the wrong they sense. How many times have public servants said that waste is not yet bad because it is not killing people yet.

The ecosystems package their biochemistry within the species each one being different and each one occupying a different pathway of processing of materials and energy. The more diversity there is in the biochemistry the more special abilities there are to process and mineralize wastes. The more complexity there is the cleaner emerge the air and waters. As long as there are complex ecosystems, man is protected. They are his diversified life support. They take the toxins first and often considerable populations are killed. So long as there is complexity remaining there is some protection for man. As the levels of waste rise, however, beyond the capacity of the life-support system, the diversity is destroyed and with it man's protection. Then the levels of waste toxicity can pass the natural front lines to kill man directly.

Man needs protection of the life-support systems diversity and this means the rare specialists—each one of which contributes to the protection of man. Funds are needed to monitor these environments for the levels of waste stress just as any other management system would monitor the adequate operation of its engines. Putting all the funds into studies of human toxicity in tissues and foods, and learning how the toxins kill us is putting the funds one step too late. By the time direct human toxicities are involved, the life-support system is in sad shape and our survival in doubt. The sensitive species are our early warning systems.

CONSTITUTIONAL RIGHT TO LIFE SUPPORT

Basic to many of the legal battles underway and developing in the defense of the environment is a long ignored constitutional freedom—the human right to a safe life-support system. There can be no more fundamental right to an individual than his opportunity to breathe, drink water, eat, and move about with safety. Long taken for granted, these rights are not free but are paid for daily by the metabolic works of the life-support system processing the wastes and by-products. The water and mineral cycles, the complex of complicated organisms that process varied chemicals, and the panorama of ecological subsystems that organize and manage the earth's surface are not the property of individuals, but are part of the essential basic right, the life-support system. A fundamental flaw in the legal systems allowed owners of land to assume special rights to the public life support means.

The situation is no longer one of seemingly unlimited reserves of environment. Now any action to foul the life-support system threatens the life of individuals everywhere. As cancer-inducing chemicals become more concentrated and widely distributed, the threat of stress makes formerly private activities into public menaces. A riparian owner who wastes or pollutes water has contributed along with many others like him to taking the right of life from people who have never seen him. Where a group contributes to murder, any individual is accessory who participates in a small way. Is the situation comparable?

Consider some magnitudes. Although the figure is probably too small, the world life-support productivity has been estimated as 1 gram of oxygen per square meter per day. This is the normal processing of non-toxic organic wastes and with it binding and removal of many wastes that are incorporated in the biological processes and stored or regenerated in "clean" form. With a world population of humans of 3.479×10^9 people and a world surface area of 5.98×10^{14} m^2, we can divide to obtain 171,000 m^2 per person (42.5 acres). Each human's portion of the earth's life-support system is 1.7×10^5 g oxygen or 4 times this amount of energy

$(6.8 \times 10^5$ kcal) processing by the system daily. Every time someone discharges about 380 pounds of organic waste per day he has diverted the life-support fraction of one person. If the substances are toxic, the amplifier destructive action is much greater. Large storages of oxygen and carbon in air and sea protect us from immediate difficulty with them, but we use their flows as an index to our disturbance of nature.

The basic right to life (to ecosystem joiners) requires revision of laws to convict anyone of willful release of waste which disorders the general life-support system, the mineral cycles, and the complex ecological systems that keep the life-support system working efficiently and smoothly. Economic developers selfishly interested in their industry, their town, or their records of expansion must include in the costs of any new ventures the full interfacing of the activity with the life-support systems and payments in full to the public sector for any energy values heretofore regarded as free (except for processing cost) such as water, air, mineral reserves, and ecosystem area. The developer may protest that it will block development. Why should there be a God-given right to develop? The right to ownership of land does not include the right to take it out of the life-support system without due process.

Once the basic facts of planetary existence are understood, the laws and court actions can protect the individual rights and through this principle incorporate nature and man in the same system of right. Probably a constitutional amendment is required within the United States. For world agreements new legal mechanisms are needed.

The levels of fuel consumption become matters of public safety since all the other overloads follow from high power flowing in the system. As soon as the race for competitive exclusion can be stopped, the right to inject fuel into the overheated world economy must be governed. The safety of the citizens of the world requires that the oils be kept in the ground and released at a rate consistent with long-range public safety. Injunctions by citizens groups against oil companies expanding fuel consumption may be in order soon. Advertisements to accelerate power consumption by power companies and proclamations by governors and state development agencies for expansion are menacing to public safety. Oil depletion allowances will be clearly unconstitutional on the ground of threat to public safety as the basic cause of overpopulations and overpollution.

POWER DENSITY

A useful principle for planning which emerges from these energy arguments is the regulated power density. As a general basic planning right, no power density [kcal/(m²)(day)] should be planned that is higher than the life-support system in which it is imbedded. Permits and laws should be required for those high-power densities that are essential to public survival. The burden of proof should be on the developer. See Table 2.2 which suggests limits depending on multiplier effects.

SUMMARY

The developing system of man and nature may be idealized as three sectors, each a full-fledged member of spaceship earth—the urban sector, the agricultural sector, and the wilderness life-support sector (Fig. 10–9). All three are equal partners in the new nature, and management of the wilderness sector is really a problem of incorporating this nonhuman part into the economy, into all plans for balancing flows, into our system of payments for service, into our sense of rights, into our laws, and into our destiny. This is done by using energy values for evaluations rather than moneys that belong to only part of the system. Special attention is needed to develop reward loops to the natural sector, interface mechanisms between sectors, and a basic change to an ethic of limited fuel management.

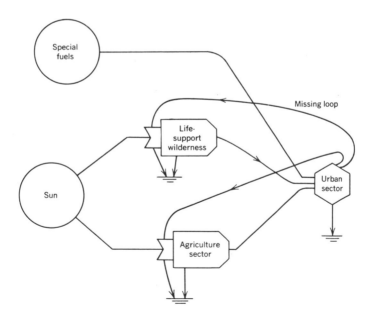

Figure 10–9 Three main sectors of industrialized economy showing the main missing loop to couple the human sectors to the life-support system of the new wilderness.

REFERENCES

[1] Anderson, A., *Marine Resources of the Corpus Christi Area*, Bureau of Business Research, University of Texas, Research Monograph No. 21, 1960.

[2] Armstrong, N., and H. T. Odum, "A Photoelectric Ecosystem," *Science*, 143, 256–258 (1963).

[3] Beyers, R. J., "Relationship between Temperature and Metabolism of Experimental Ecosystems," *Science*, 136, 980–982 (1962).

[4] Bradshaw, A. D., T. S. McNeilly, and R. P. G. Gregory, "Industrialization, Evolution and Development of Heavy Metal Tolerance in Plants," in *British Ecological Society Symposium Number 5*, John Wiley and Sons, New York, 1965, pp. 327–343.

[5] Cooke, G. D., R. J. Beyers, and E. P. Odum, "The Case for the Multispecies Ecological System with Special Reference to Succession and Stability," in *Bioregenerative Systems*, J. F. Saunders (ed.), National Aeronautics and Space Administration, NASA SP-165, 1966, pp. 129–139.

[6] Elton, C. S., *The Ecology of Invasions by Animals and Plants*, Methuen, London, 1958.

[7] Murphy, E. F., *Governing Nature*, Quadrangle Books, Chicago, Ill., 1967.

[8] Nelson, L. W., and R. Ewell, "Manufactured Physical and Biological Inputs," in *The World Food Problem*, Vol. 3, Report of President's Science Advisory Committee Panel on World Food Supply, White House, Washington, D.C., 1967, pp. 95–216.

[9] O'Brien, R. D., *Insecticides*, Academic Press, New York, 1967.

[10] Odum, E. P. with collaboration of Howard T. Odum, *Fundamentals of Ecology*, W. B. Saunders Co., Philadelphia, 1959.

[11] Odum, H. T. and R. C. Pinkerton, "Time's Speed Regulator, the Optimum Efficiency for Maximum Output in Physical and Biological Systems," *Amer. Sci.*, 43, 331–343 (1955).

[12] Odum, E. P., "Relationship between Structure and Function in the Ecosystem," *Japanese Ecol.*, 12, 108–118 (1962).

[13] Odum, H. T., W. L. Siler, R. L. Beyers, and N. Armstrong, "Experiments with Engineering of Marine Ecosystems," *Publ. Inst. Mar. Sci. Univ. Texas*, 9, 373–453 (1963).

[14] Odum, H. T., "Energetics of World Food Production," in *The World Food Problems*, Vol. 3, Report of President's Science Advisory Committee Panel on World Food Supply, White House, Washington, D.C., 1967, pp. 55–94.

[15] Odum, H. T., "Limits of Remote Ecosystems Containing Man," *Amer. Biology Teacher*, 25, 429–443.

[16] Odum, H. T. and R. F. Pigeon (eds.), *A Tropical Rain Forest*, Division of Technical Information, U.S. Atomic Energy Commission, Oak Ridge, Tennessee, 1970. See Chapter I-3 by Odum, Lugo, and Burns.

[17] Olson, J. S., "Energy Storage and the Balance of Producers and Decomposers in Ecological Systems," *Ecology*, 44, 322–331 (1963).

[18] Owen, D. F., *Animal Ecology in Tropical Africa*, W. H. Freeman and Co., San Francisco, 1966.

[19] President's Science Advisory Committee, "Tropical Soils and Climates," in *The World Food Problem*, Vol. II, White House, Washington, D.C. 1967, pp. 471–500.

[20] Putnam, P. C., *Energy in the Future*, D. Van Nostrand, Princeton, N.J., 1953.

[21] Saunders, J. F., *Bioregenerative Systems*, National Aeronautics and Space Administration, NASA SP–165 (1966).

[22] Von Bertalanffy, L., *General System Theory*, George Braziller, New York, 1968.

11

WHAT SYSTEMS ARE NEXT?

We may summarize the story of man and nature by looking ahead. The future can be divided into alternatives according to energy supplies. There is the *future of power expanding*, the *future of power constant*, and *the future of power receding*. National and international planning task forces should be assigned to each of the three contingencies.

We are not now sure which future will be next; perhaps they will follow each other in a step-by-step sequence. These energy alternatives concern survival and should probably demand far more of our national attention than the status symbol of space travel. The availability of enough energy for any significant amount of space activity is simply not in sight. The space program may indeed be a lovely flower in the late summer of man's current growing season. We have energy now to plan for the future, but later there may be no excess calories.

FUTURE OF POWER EXPANDING

If and as long as sustained and increasing sources of potential energy are developed from fossil fuels and nuclear power, there will be advancing patterns of self-design and ever more complex networks of man. If an awareness of energy laws can be better dispersed among the world's citizens, some improved designs of political structure may yet develop a one-world system, bypassing military power testing. For a one-world structure, however, the science of energy networks in sociosystems must be urgently advanced. For survival we may need such national projects as the development of a giant passive analog network model for simulating man's system of energy and economics in detail.

A great danger may be incomplete looping of materials. Wastes may accumulate, and the minerals may fail to reach enough agriculture. If the world designs do not connect the wastes regeneratively to the food input, both starvation and pollution result as the two faces of the same difficulty in systems design. It is also a hazard that self-designing readjustments occur only after levels of chemical content toxic to the biosphere develop, excluding man locally and possibly even completely. For the engineering of better regenerative ecosystems consistent with the new inputs of civilized man, a national project (see Fig. 10–3) is needed on ecological engineering on a large scale, hopefully to develop new combinations of species that can deal with the special chemistry involved.

During a future of power expanding, some new means will be required in the educational realm to incorporate all the problems of accelerating complexity and provide sensible roles and perspectives for each citizen. The science of general systems may help to condense knowledge and provide overviews. Perhaps systems science will be introduced in the grammar school, replacing the miscellaneous materials of general science courses. Much information now taught in dozens of graduate specialties involves different special cases of the same network properties; this information could be organized in a unified curriculum based on systems principles, perhaps simplifying education. The student might then again see where he stands in existence. In the process of reorganizing education around systems principles, a general science of the noosystem[1] may develop to explain the phenomena of man and the ecosystem. Ecology may move closer to sociology and the other social sciences.

Many specialties such as oceanography, limnology, forestry, agronomy, and wildlife and fisheries management may form a common discipline, not so far removed from the present social sciences. New systems approaches may be reflected in new campus divisional groups streamlining educational curricula around the size levels of organizations through which energies flow: the atomic, the molecular, the biological, and the ecological (in its broader sense). With more specialities, however, many more permutations of learning are possible and desirable, and curricula must be less rigid to allow new combinations.

There is the fearful prospect that overpopulation will spread from its underdeveloped focuses to encompass the world instead of falling back

---

[1] Some new words may be needed for the study of systems of man and nature. "Ecosystem" might be expanded in its applicability. The definition of the noosphere implies the sociosystem and the ecosystem.

before the spread of more enlightened systems. The dangers are somewhat clarified by the energy concepts. A country that limits its population is not threatened by one that overpopulates as long as its total power budget is in excess and it channels its energies to include reserves and adequate defense. The country that does not limit its ratio of population to energy budget puts its energies into low-level maintenance of ignorance, regimentation, and the many miseries that are the aftermaths of following such a course. Unfortunately, these countries can focus great power in endeavors such as military experimentation, although the power focused cannot exceed that possible from countries with greater energy budgets. An important unsolved question for simulation is the nature of the world system that would be imposed on its members by a world government encompassing countries with two energy systems, the one dedicated to a maximum population-to-energy-budget ratio, the other to a minimum-population-to-energy-budget ratio.

POWER CONTROL AT THE MINE AND WELLHEAD

The greatest need of all, largely unrealized in a world that talks now directly of population explosion and economic runaway is to gain control of the basic cause, which is the accelerating outflow of potential energy from the fossil-fuel supplies. In some way the control of this flow must be placed in the public sector, for the public good, and its flow stabilized. There must be enough international agreement to prevent runaway efforts of one country bent on competitive exponential growth which seems temporarily good for them but cancerous for all in the long run. This is a fearsome task with most of the fuel supplies in the hands of small partly developed countries and oil companies both of which are under short-range planning only. We have a chance if first we can explain to the citizens of the world the underlying energy causes of our situation. Then the energy reforms of international law can follow as indicated in Chapter 7. This will require incorporation of the energy ethic in all the religious programs as indicated in Chapter 8. If as indicated in Chapter 6 our structure of economists advising our planning endeavors will incorporate the basic energy drives in energy-econometric models, then we may substitute more certainty for the heretofore elusive ability to predict economic details using an incomplete model. The rising levels of education and one-world television communication may make all this possible.

If it is correct that use of potential energy tends to develop equal distributions in a network, gradual equalizations may cut down the forces thrust across network junctions between poor and rich areas. The stability of dominant protectors may be replaced by some stability of institutional unification in order to focus enough power for world organization.

A decline in expansion may allow many aspects of evolution to catch up in self-design processes. The regenerative system of nature may evolve the ecosystems for dealing with man's waste and recirculating the materials back into agriculture. The airs and waters may again be clear.

With leveling of energies, the hazards of energy shortage begin to develop in hidden ways. Development of more network structure and new programs requiring maintenance may use the energy reserves and margins. Then, if fluctuations and demands for power delivery are forced upon the system, and if controls are not specifically designed to prevent it, energies may be diverted from the organizational institutions rather than from individuals. Once organization falters, the components may separate with no energy focus available to reunite them. Safety may require a religious switching system in individuals able to make them put the system first under this kind of stress, since the ultimate safety of the individual is in the stable system. Highly individualistic priorities and loyalties may be dangerous in complex states that have no energy excess.

FUTURE OF POWER RECEDING

The fossil record is full of systems that rose and fell to extinction. Biological specialization in organic body development tends to be one-way, new developments coming from the undifferentiated, unspecialized cells that are tucked away amidst the powerful main structures of action like muscle and nerve. In analogy with these facts, many authors after Hegel have suggested the dangers of extinction of the main civilization that might follow any effort to decrease its activity. Can complex civilization de-differentiate? Are generalized parts such as the youth a means for programming change? Certainly it is an important contingency to consider the possible future of civilization if and as the energy budgets recede back toward the level of energy available from solar income.

We may eliminate fear of necessary extinction and disintegration by pointing to the orderly rise and fall of many ecosystems in rhythm with the rise and fall of the seasonal energy budgets. Systems such as some temperate forests, the Atlantic estuaries, and the Silver Springs River experience a tenfold or greater change in their energy flows each year without destruction. These systems have mechanisms for reducing their drains, diminishing their populations, dispersing concentrations, arranging migrations, and maintaining their general order in a more dilute form consistent with the diminished energy budget. The challenge to any national task force assigned the responsibility consists of preparing for the contingency of declining power to develop such a transition plan for man. The pattern need not be sudden collapse, although precedent exists for both adjustment and catastrophe.

While energies are still in excess, adequate preparations can be made for preserving and holding the needed knowledge and cultural memory in libraries and institutions. Then plans can be made for a more agrarian system, benefited by the knowledge we now have about them. We can plan for smaller cities, fewer cars, greater ratios of agricultural workers to town consumers, and fewer problems with pollution.

FUTURE OF POWER CONSTANT

Perhaps the fossil-fuel-based energy explosion is not long for this world. The curve of exponentially rising energy use may be crossing the curve of increasing costs of finding cheap new energy sources. As the ratio of potential energy found to work expended starts to decline, the activities involving energy excesses may disappear; then the amount of structure and useful function that can be supported will stop increasing. There may be a long period of leveling energy budgets of a fairly high plane, but the expanding economy may be gone. The citizen will sense this process as inflation.

As for religion, a stabilization of change may allow the codification of yet unfound faiths that characterize the now new networks. Possibly new prophets will emerge for fresh religious adaptations to the complex modern systems. An adapted religion may cause great energy economies in the switching systems that dedicate people to group causes ultimately favorable to themselves. See commandments of the energy ethic (Table 8–1).

Stability allows complex diversity and uniqueness of individuals. As in Augustinian Rome, there may be golden eras—if men can be satisfied with small causes, for energies big enough for new causes would have to be diverted from older endeavors. This in turn requires a willingness to discard activities.

Population can be drastically cut back at any time by crash efforts at birth control in proportion to expected energy budget; such a cutback is possibly essential to prevent collapse. In this adjustment man will be following the thousands of precedents in long-surviving populations of animals.

There will be a fine heritage of institutions and customs that cost relatively little to keep in diluted function, even though their too spacious buildings may have to receive a lower level of maintenance. The lesser era may hold a better organization than Europe's Middle Ages did in relation to Rome. As long as the energy ratio to the individual is high, the standard of living need not be low, and the air and waters will be beautiful again. If the recession is carried out slowly with the critical aspects controlled by plan, the world might well be better off and life quite livable. The field that is now called history will thrive in a new renaissance, providing guiding plans as nearly forgotten mechanisms of lower-energy societies are returned to function after centuries of disuse and relegation to museums and remote cultural byways.

A bright possibility is *ecological engineering*. Adequate knowledge about the natural solar-energy-based system may allow a small concentrated loopback of energy to guide the systems of fields, forests, and seas to stabilize and produce for man. Although there is yet excess energy, it might be better to put crash efforts into ecological engineering rather than into space. A knowledge of natural system control will be of vastly greater survival value to man than a memory of space exploration. As shown in previous chapters, only a small percentage of energies taken from natural systems need be put back at their upstream gate control circuits to direct them toward the unit chosen to receive the yield.

We do not know what role diversity should take in the stability of systems where yields are needed. Perhaps a Swedish pattern will prevail in which there is a diversity of yield systems in small alternating patches, each one being uniform. For example, there are spruce plantings for paper, intensified agriculture ponds, and urban zones. Some natural marine sea bottoms also have diversity in plots, each of which is uniform within.

The social systems must also be prepared. Low-energy states will make war less terrible than now, and if knowledge of the ways to organize against competitive exclusion tendencies can be developed while we still have excess energy, wars may be eliminated. As with ecological planning there is the same need to use available excess energies to plan for future system control of a low-energy network of man. Loopback focus of relatively small energies may be able to hold world organization. If religious systems can develop a network-favoring ethic as a switching system to hold individuals in support of world order, the manifold organization set up at high energy levels may be maintained at later low energy levels. Greater work expenditures are required to create novel structure than to maintain it. The preparations need to be made now for this contingency of receding power.

THE NEW PROPHECY

These three contingencies must practice the same energy economy in adapting individuals to the new ideals based on new systems of man. While there is energy, we need to stimulate religious evolution.

We may encourage faster religious change even now by injecting large doses of systems science into the training of religious leaders. What a glorious flood of new revelation of truth God (the essence of network) has handed man in the twentieth century through sciences and other creative endeavors. How false are the prophets who refuse even to read about them and interpret the message to the flock. Why do some inhabitants of the church pulpits fight the new revelations simply because the temporary prophets are a million spiritually humble little people in laboratories and libraries, only vaguely aware of their role? Why not open the church doors to the new religion and use the preadapted cathedrals and best ethics of the old to include the new? Let us inject systems science in overdoses into the seminaries and see what happens. Why should we fear that deviation from rigid symbols of the old religion is deviation from morality? A new and more powerful morality may emerge through the dedication of the millions of men who have faith in the new networks and endeavor zealously for them. Prophet where art thou?

APPENDIX: ENERGY MODULE FORMULAS

Each of the energy modules refers to a class of related components and processes in the systems of energy flow as already described in connection with Figure 2–4. For each module there are equations that describe the operation of the module in its simplest form, that is, when it has no special complications. For example, the self-maintenance module (Fig. 2–4(g)) with one feedback work loop has properties of logistic growth with an equation given in Figure 11–4, but the usual module in the real world has more than one work loop often with additional control pathways such as the example diagrammed in Figures 3–13 and 5–1. The system of equations describing the module's operation then becomes more complex, although the S-shape growth may serve still as a mental model for qualitative considerations, at least until such time as the more complex network is actually simulated.

These equations will remind various scientific and engineering readers of familiar equations in their fields of study and experience and serve to help them connect their knowledge to these symbols. A discussion of each symbol in relation to energy and systems concepts is given elsewhere [Odum, H. T. An Energy Circuit Language for Ecological and Social Systems, *in Systems Analysis and Simulation in Ecology* ed. by B. Patton, Academic Press.], and some illustrations from that source are provided as Figures 11–1 through 11–5.

The expressions that accompany each module are as follows.

Energy Source (Fig. 2–4(a))

Some of the following occur in various situations.

$$X = k \qquad \text{Flow constant}$$
$$J = k \qquad \text{Force constant}$$
$$X = kt \qquad \text{Force or flow as ramp}$$
$$j = kt$$
$$X = \sin kt \qquad \text{Sine function}$$

Passive Storage (Fig. 2–4(b))

Module follows Von Bertalanffy exponential growth, called a first-order process in engineering; see various ways of expression in Figure 11–1.

Heat Sink (Fig. 2–4(c))

Module described by rate of entropy(s) increase from rate of potential energy decline dispersing heat(w).

$$\frac{dF}{dt} = \frac{dw}{dt} = \frac{TdS}{dt}$$

Potential-Generating Work (Fig. 2–4(d))

Module stores energy in relation to that released into heat according to loading of forces shown in Figure 11–2. When loaded for maximum power storage, the force loading is 50 percent.

Work Gate (Fig. 2–4(f))

Module in its simplest form has an output that is the product of the local input forces at the site as shown in Figure 2–6. The expressions for the response with one limited is given in Figure 11–3.

Cycling Receptor (Fig. 2–4(e))

Module is the general action of which the Michaelis-Menton formulation was the first application as given in Figure 11–5.

Self-Maintenance (Fig. 2–4(g))

Module in simplest form has the following expressions based on Figure 11–4(a) with backforces important. The growth with time from initial zero start has an S-shape growth (Fig. 5–8).

$$J_q = J_o = \frac{L \ (N_1 - N_2)}{2} \qquad L = k_o J_2$$
$$J_2 = BN_2$$
$$J_3 = kN_2$$

$$\frac{dQ}{dt} = J_q - J_2 - J_3$$

$$\frac{dQ}{dt} = \frac{1}{C} \left[\frac{k_oBN_1}{2} - B - k \right] Q - \frac{k_oB}{2C^2} Q^2$$

Figure 11–4(b), even without backforces operating, has a square negative term and follows the logistic growth curve. Figure 11–4(c) has similar form but with more positive and negative terms due to extra feedback multiplications, and Figure 11–4(e) has a programed time lag in the feedback which gives the module oscillator or damping aspects.

Green Plant (Fig. 2–4(h))

The combination modules from a cycling receptor and one or more self-maintenance modules is the combination of the system of the separate equations (see simplest example in Fig. 9–4).

Switch (Fig. 2–4(i))

Module refers to any action which has on or off positions controlled by some code of input combinations which are above or below thresholds according to the various combinations of digital logic which must be specified for full description (AND, OR, NAND, NOR, Exclusive OR, etc.).

Transaction (Fig. 2–4(j))

Module refers to counter-current control. j_1 is power flow; j_2 is dollar flow.

$$j_1 = k_{21}j_2$$
$$j_2 = k_{12}j_1$$

Constant Gain Amplifier (Fig. 2–4(k))

Module has same function used in electrical engineering for operational amplifiers except as used here there is no automatic change in algebraic sign (G is gain).

Acceleration Impedance (Fig. 2–4(l))

Module in its simplest form has the response characteristics of an electrical inductance (see texts on electrical engineering). Backforce is proportional to acceleration.

One-Way Valve (Fig. 2–4(m))

Module has response of electrical diode with flow proportional to force, with backforces expressed, but no back flow.

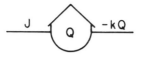

Energy

$$\frac{dQ}{dt} = J - kQ$$ Differential

$$Q = \int (J - kQ)\, dt$$ Integral

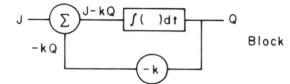

Block

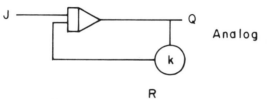

Analog

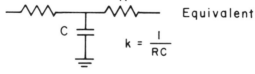

Equivalent

$$Q = JRC\left(1 - e^{-\frac{t}{RC}}\right)$$ Integrated

Figure 11-1 Performance characteristics of the Von Bertalanffy module given in several languages.

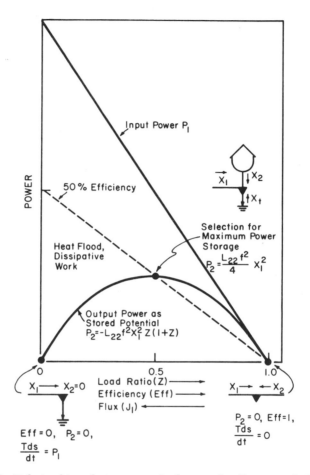

Figure 11–2 Relationship of power to loading and efficiency of the potential generating storage module.

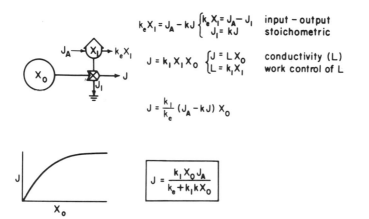

$$k_e X_I = J_A - kJ \begin{cases} k_e X_I = J_A - J_I & \text{input – output} \\ J_I = kJ & \text{stoichometric} \end{cases}$$

$$J = k_I X_I X_O \begin{cases} J = L X_O & \text{conductivity (L)} \\ L = k_I X_I & \text{work control of L} \end{cases}$$

$$J = \frac{k_I}{k_e} (J_A - kJ) X_O$$

$$\boxed{J = \frac{k_I X_O J_A}{k_e + k_I k X_O}}$$

Figure 11–3 Performance characteristics of the work gate when one input flow is constant and limiting.

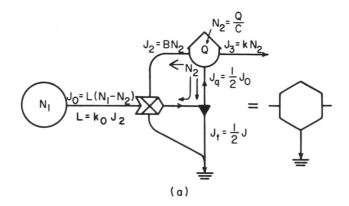

(a)

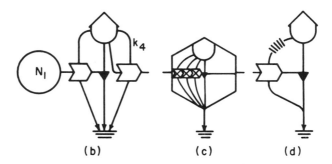

(b)   (c)   (d)

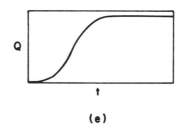

(e)

Figure 11–4 Self-maintaining module in its simpler forms. System of relationships in (a) is logistic; (b) quadratic outflow pathway; (c) 3 input reward loop work gates; (d) self maintaining work pathway with a time lag impedance; (e) "S" shaped curve characteristic of the logistic and related growth forms exhibited by some of these modules.

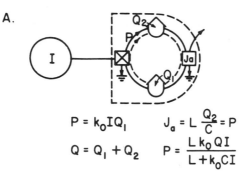

A.

$$P = k_0 I Q_1 \qquad J_a = L \frac{Q_2}{C} = P$$

$$Q = Q_1 + Q_2 \qquad P = \frac{L k_0 Q I}{L + k_0 C I}$$

equals:

B.

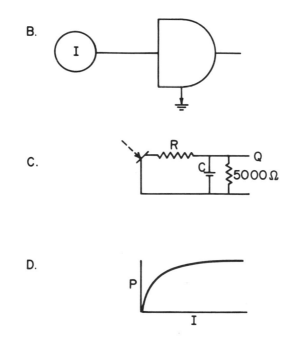

C.

D.

Figure 11–5 Performance characteristics of the cycling receptor module: (*a*) component pathways and relationships implied in the basic modular symbol (*b*). (*c*) Passive analog using photometer cell, a special case for simulation; (*d*) output response with varying input intensity (*I*).

INDEX